SpringerBriefs in Physics

W0036988

More information about this series at http://www.springer.com/series/8902

Andrea Puglisi

Transport and Fluctuations in Granular Fluids

From Boltzmann Equation
to Hydrodynamics, Diffusion
and Motor Effects

 Springer

Andrea Puglisi
Istituto dei Sistemi Complessi
Consiglio Nazionale delle Ricerche
Rome
Italy

and

CNR-ISC c/o Dipartimento di Fisica
Sapienza-Università di Roma
Rome
Italy

ISSN 2191-5423   ISSN 2191-5431   (electronic)
ISBN 978-3-319-10285-6   ISBN 978-3-319-10286-3   (eBook)
DOI 10.1007/978-3-319-10286-3

Library of Congress Control Number: 2014947648

Springer Cham Heidelberg New York Dordrecht London

Printed on acid-free paper

Springer is part of Springer Science+Business Media (www.springer.com)

A Fabiana e Luca

Preface

The purpose of the book is to discuss nonequilibrium phenomena in fluidized granular materials, with an accent on granular kinetic theory and some of its stochastic extensions. A few other books exist on this subject. The difference in this new one is to provide the reader with a *brief* introduction, which goes through a few salient points in the subject: models for collisions, Boltzmann equation, fundamental boundary conditions, transport equations and hydrodynamics, macroscopic ordered phenomena, the motion of tracer particles, and the breaking of time-reversal symmetry.

This book is *not* a topical review. Therefore reference sections are not meant to be exhaustive. My intent is to offer a selection of starting points, for instance by citing reviews or other books, where the reader will find more detailed bibliographies. This book merges material from two courses given for Ph.D. students at the Physics Department of Sapienza University (2010 and 2012), and a general reorganization of the results of GranularChaos project. GranularChaos is a 5-year-long project (2009–2014) funded by the Italian Ministry for University and Research, after winning the selection at the Eureopean Research Council (Starting Grant 2007). The focus of the project is fluctuations in granular media.

I am indebted to Angelo Vulpiani for most of what I learned about nonequilibrium statistical mechanics and granular materials: collaboration with Angelo has always been enjoyable and fruitful. The hint to write this Brief came from him: again an interesting and challenging incitement. Umberto Marini Bettolo Marconi, Alberto Petri and Vittorio Loreto are the other three friends who greatly improved my knowledge and understanding of the subject, since the beginning of my study of granular fluids. Many of the ideas and results contained in this book are due to collaboration and discussions with them, during the last 15 years and more. I wish to say thanks also to Andrea Baldassarri, who shared with me many progresses on granular kinetic models and, in the early years of my doctorate, was an intense stimulus to become a better c-programmer and a more careful researcher. My understanding of granular fluids in the wider context of nonequilibrium steady states has received a great impulse during a stay of 2 years in Orsay (Paris), where I collaborated with Alain Barrat, Emmanuel Trizac, and Frederic van-Wijland, whom I wish to warmly thank. In the last years I had the exciting possibility, as a coordinator of the GranularChaos project, to interact with brilliant young collaborators, in particular with Giulio Costantini, Giacomo Gradenigo, Alessandro Sarracino, and Dario Villamaina, whom I acknowledge for a constant passion, curiosity, and their many intriguing interrogatives: they shaped my ability to explain and teach. My hope is that, as a consequence, this book will be clear and useful to students and young researchers. The GranularChaos project, thanks to the crucial help of Andrea Gnoli, has allowed me to enter in the fascinating world of real experiments with granular fluids, an experience which has deeply influenced my perspective on this subject. A special acknowledgment goes to Andrea Gnoli, Alessandro Sarracino, Camille Scalliet, and Angelo Vulpiani, who read the manuscript, found plenty of errors, and gave me many useful advices.

Last, but certainly not least, I wish to thank my beloved family, in particular Fabiana (well before that unstoppable and joyful creature which dwells in our house since a couple of years), for their patience, tolerance, and love.

Roma, July 2014   Andrea Puglisi

Contents

Symbols

σ	Particle's diameter
$\hat{n}$	Unit vector joining particles' centers of mass in a collision
λ	Mean free path
τ_c	Mean free time between collisions
ω_c	Mean collision frequency
S	Total scattering cross section
v_T	Thermal velocity
r	Restitution coefficient
N	Number of particles
$P(\mathbf{r}, \mathbf{v}, t)$	Single particle probability density function (normalized to 1)
$Q(P, P)$	Collisional integral in the Boltzmann equation
V_{12}	Relative velocity between colliding particles 1 and 2
g	Modulus of relative velocity between colliding particles projected along $\hat{n}$
$g_2(\sigma)$	Spatial pair correlation evaluated at contact
ϕ	Packing (or "volume") fraction
μ_p	p-th velocity moment of the collisional integral
ζ	Collisional cooling rate
a_2	Coefficient of the second polynomial in the Sonine expansion
T	Granular temperature
ε	Small parameter for perturbative expansions, e.g. Knudsen number (Chap. 3) or square root of the mass ratio (Chap. 4)
γ	Shear rate
η	Viscosity
κ	Thermal conductivity
μ	Dufour-like thermal conductivity

$\xi_\perp, \xi_\parallel, \xi_T$ Correlation lengths related to linearized hydrodynamics

ξ, ξ', ξ'' Noises in fluctuating hydrodynamics

$U_\perp$ Component of the (Fourier-transformed) velocity field perpendicular to the wave vector

$C_\perp$ Time autocorrelation of $U_\perp$

$\mathscr{E}$ Noise in Langevin equations for tracer's dynamics

Introduction and Motivation

A granular material is a substance made of grains, i.e., many macroscopic particles with a spatial extension (average diameter) that ranges from tenths of microns to millimeters. In line of principle the size of grains is not limited as far as their behavior can be described by classical mechanics. For example, the physics of planetary rings (made of objects with a diameter far larger than centimeters) is sometimes studied with models of granular media. More often the term "granular" applies to industrial powders: in chemical or pharmaceutical industries the problem of mixing or separating different kinds of powders is well known; the problem of the transport of pills, seeds, concretes, etc., is also widely studied by engineers; the prevention of avalanches or the study of formation and motion of desert dunes are the subject of important studies, often involving granular theories; silos containing granular products from agriculture sometimes undergo to dramatic breakages, or more often their content become irreversibly stuck in the inside, because of huge internal force chains; the problem of diffusion of fluids through densely packed granular materials is vital for the industry of natural combustibles; the study of ripples formations in the sand under shallow seawaters can solve important emergencies on many coasts of the world. Rough estimates of the losses suffered in the world economy due to ignorance of granular *laws* amount to billions of dollars a year.

The study of granular materials dates back to the nineteenth century, with the first studies by Coulomb in 1773, Faraday in 1831, Reynolds in 1885 and much more recently by Bagnold, who really opened the way to the systematic study of granular rheology. For decades granular systems have been a blessing and a curse for engineers. In the last 50 years they have become more and more present in physics laboratories. The rise of computer simulations has led to a huge increase of interest in the study of realistic granular models.

In parallel, a closer look at the fundamental properties of granular media (inelasticity of collisions and entropic constraints) motivated the introduction of new minimal models. These "granular cartoons" have the remarkable charm of displaying an intriguing behavior in spite of their simplicity. Granular gases represent a noteworthy example. As for spin glasses, some models of granular gas can be observed only in the silicon cage of a computer simulation. Nevertheless, their study is fundamental to understand the relevance of the basic assumptions (and limits) of Kinetic Theory, Hydrodynamics, and general nonequilibrium Statistical Mechanics.

The goal of this book is to present a brief sketch of the many theoretical tools and models which succeed in describing fundamental granular experiments. Structures, patterns, correlations, and motor effects, all phenomena that cannot appear in equilibrium molecular fluids, are found to be common in flowing granular materials. Simple microscopic models based on inelastic collisions reproduce all these effects. At the same time they constitute a possible starting point for the developement of granular hydrodynamics. Such a macroscopic theory for "slow" granular flow is hopefully more robust than its microscopic foundation. Frequently in this book the reader will find examples of such a counterintuitive effect. Starting from microscopic Molecular Chaos, macroscopic ordered structures, correlations, clusters, convective cells, or directed motion will appear. This is not a contradiction of the hypothesis: on the contrary, it is the beauty of new phenomena emerging when the scales of description change.

The purpose of Chap. 1 is to offer first a quick introduction to the "wild world" of granular materials, which goes well beyond the study of granular fluids, and then a more focused overview of phenomena observed in fluidized granular media. In Chap. 2 I sketch the classical derivation of Boltzmann kinetic equation from the Liouville equation, discussing the points where a granular gas differs from a molecular one. In Chap. 3 I review the basic steps of the Chapman-Enskog procedure to derive hydrodynamics from the Boltzmann equation. In this chapter I also discuss the arguments given by L. Kadanoff and I. Goldhirsch who criticized the blind application of hydrodynamics to granular fluids, and I conclude with some noteworthy applications where hydrodynamics gives a fair description of observed phenomena. In Chap. 4 the diffusion of a tracer is discussed in different limits, and the case of an asymmetric tracer is given as an example of "granular ratchet," also realized in recent experiments. In Chap. 5, finally, I revisit some of the models introduced in the previous chapters, in the broader perspective of nonequilibrium statistical mechanics and stochastic processes, by discussing linear response and entropy production.

Chapter 1
Granular Fluids: From Everyday Life to the Lab

Abstract In this chapter, I introduce the basic concepts and tools useful to study granular media. A tour is offered through some of the many fascinating granular phenomena. These include hydrodynamic instabilities such as granular jets, fingering, spontaneous segregation, thermal-like convection, and the several ratchet-like phenomena where "thermal" fluctuations are somehow *rectified*. The analogy with active fluids is also discussed.

1.1 The Granular "States"

Physicists try to reduce the complexity of real situations. Such an attitude toward simplification is evident in experiments where the fundamental behavior of granular media is probed. The models proposed by theoretical physicists are even more idealized, in order to catch the essential ingredients of phenomena. In an experiment the grains are often smooth spheres with the same size, same restitution coefficient, perfectly dry, possibly in microgravity and in the void, and so on. In a numerical simulation the grains can become rods moving on a segment or disks with constant restitution coefficient. The effect of a shaker can be modelled as a thermostat. Notwithstanding such a multiplicity of viewpoints, some fundamental ingredients are common in all the approaches and, in a sense, constitute the definition, in physics, of the granular state of matter.

Granular matter is distinct from the usual molecular matter because of the size of elementary constituents. Grains are macroscopic, typically with a minimum linear size of ~ 0.1 mm. As a consequence, they are described by rules of classical mechanics with dissipative interactions. In a collision the kinetic energy of grains' centers of masses is transferred to internal degrees of freedom, i.e. heat, and rapidly dispersed to the environment. In essence, a fraction of kinetic energy disappears from the description of the system. Dissipative interactions have many implications, the most fundamental being the breakdown of symmetry under time-reversal. Furthermore, the mass of a grain is of the order of 10^{20} molecular masses: the kinetic or potential energy of a grain is therefore many orders of magnitude larger than molecular thermal energy. As a matter of fact, the temperature of the environment plays a negligible role in the dynamics of the grains, i.e. they can be safely considered at $T = 0$.

© The Author(s) 2015
A. Puglisi, *Transport and Fluctuations in Granular Fluids*,
SpringerBriefs in Physics, DOI 10.1007/978-3-319-10286-3_1

In the kinetic theory of granular gases the role of "microscopic degrees" is played by the grains themselves, and a "granular temperature" is introduced in terms of the kinetic energy of grains. To realize a motion of the granular particles, some kind of "thermostat" is required. Several ways exist to inject energy into a granular system: the most used is by applying forces to the container, i.e. by moving it. The motion of the container is transferred to the grains, through grain-boundary collisions. Moreover grains can be considered rigid bodies for many purposes: the volume occupied by a grain is excluded by the volume available for all the other grains. When total occupied volume is a relevant fraction of the total available one, this property has important consequences: geometrical frustration, strong spatial correlations, relevance of collisional transport versus streaming transport, enhancement of re-collisions in the kinetic equations (breakdown of molecular chaos), and much more.

In view of the above mentioned *essential features* of granular matter, it is customary to take as a reference two opposite "limit states": granular solids and granular gases [35, 49]. A real granular material is usually in an intermediate state between these two limits, depending upon the external conditions, available volume, intensity of the driving, degree of inelasticity, presence of interstitial fluids, and so on. The experimental and theoretical instruments used to tackle granular solids, for instance elastoplastic continuum models, can be very different from those applied to the study of granular gases, where one typically resorts to kinetic theory.

A few main categories of granular problems can be individuated in the literature of the last 30 years.

- Stable or metastable granular systems: this family comprehends the study of the distribution and the analysis of correlations of the internal forces in a pile or silo of grains, the characterization of the propagation of sound inside densely packed arrays, the very slow compaction dynamics observed under tapping (the grains can rest in a metastable state, in the absence of vibration, which is far from the minimum packing fraction attainable), the study of time and size distributions of avalanches in a pile which has reached its critical slope [13].
- Slow granular flows: within this regime, particles stay in contact and interact frictionally with their neighbors over long periods of time. This is the "quasi-static" regime of granular flow and is typically studied using modified plasticity models based on a Coulomb friction criterion [1].
- Rapid-flow regime: this corresponds to high-speed flows [7, 80]. Instead of moving in many-particle blocks, each particle typically moves freely and "independently" from the others. In the rapid-flow regime, the velocity of each particle may be decomposed into a sum of the mean velocity of the bulk material and an apparently random component to describe the motion of the particle relative to the mean. The analogy between the random motion of the granular particles and the thermal motion of molecules in the kinetic-theory picture of gases is strong: building upon such an analogy, the mean-square value of the random velocities is commonly referred to as the "granular temperature"—a term first used by Ogawa [63]. When the stationary velocity of the flow increases (due to an increase of external driving forces) the shear work induced by internal friction generates granular temperature

and granular pressure, which in exchange produces a decrease of volume fraction occupied by grains [7]. This suggests that a rapid flow is likely to be dilute and that theoretical methods belonging to kinetic theory, as well as a hydrodynamic description, can be tried and are sometimes successful. Every kind of typical fluid experiment has been performed on granular systems: from Couette cells to inclined channels to rotating drums, finding non-linear constitutive relations. High amplitude vibrations can generate interesting convection phenomena, associated to size and density segregation. Patterns, such as two dimensional standing waves, can form on the free surface of a vibrated granular layer. The study of simulated models posed new questions on the constitutive behavior in rapid flows. Recent experiments and numerical studies have focused on this subject, measuring the velocity probability distribution functions and finding that in a wide set of situations this distribution is not Gaussian. The study of internal stress fluctuations and of velocity structure factors has given further elements to adjust granular kinetic theories. A debate has developed on the limits of application of hydrodynamic formalism, see Sect. 3.2.

1.2 Granular Flows

This section briefly reviews a few common situations where the granular materials behave as fluids. There is not a unique classification of experiments for granular fluids: the number of different setups investigated is quite large and the number of observed phenomena is even much larger. My personal choice is that of grouping experiments on the basis of the mechanism of energy injection. This is the part of each apparatus which is better controlled by the experimentalists.

1.2.1 Air Fluidization

Air or other gases can be continuosly pumped through a container filled by grain. The result is a state of granular fluidization that can be controlled by pressure or velocity of the injected gas. The method is inspired to real applications, e.g. in mineral or metallurgical engineering, where it is called *fluidized bed technology*. It is effective in creating a fast and uniform granular flow. Recent experiments with air fluidization have demonstrated the potential of such a technique to probe fundamental features of granular dynamics. In a horizontal quasi-2d setup, for instance, an upward flow of air induce stochastic motion of grains on a plate. Within this setup, spatial heterogeneity increasing with packing fraction has been investigated. A strong analogy has been found with structural glasses near the glass transition, where long time-scales are associated with large spatial correlation lengths [39].

An interesting question arises in the context of air fluidization, concerning the role of surface-tension effects. A Hele-Shaw cell is used for this purpose: the cell is constituted by two plates separated by a gap of a few (2 or 3) grain's diameters. The gap is filled with grains and a hole in the middle of the cell lets a gas be pumped into the system at fixed high pressure. The gas expands through the packed granular material, creating a *fingering* pattern whose dynamical and geometrical properties reveal features of the granular fluid. The same phenomenon is well known when the experiment is performed with two fluids of different viscosity, where the width of the fingers is related to the capillary number, that is the ratio between viscous drag and surface tension. Usually fingers are larger and smoother as the velocity of the low viscous fluid decreases. The granular fingering experiment [9] shows the opposite behavior, i.e. fingers' width *increases* with the fluid velocity at the granular-fluid interface. Moreover, the fractal dimension of patterns is close to that of the diffusion-limited-aggregation (DLA) model, which is expected for fluids in the limit of zero surface tension.

Another instability of fluids explained by hydrodynamics is the Rayleigh instability. This is usually observed when a free falling stream of fluid breaks up into smaller packets with the same volume but less surface area, exploited also in ink-jet technology. Again the driving force of the instability is surface tension. Experiments have been performed [76] by following with a fast camera a falling granular stream. The fascinating break-up into droplet patterns is reproduced by certain granular materials and not by others. A debate about the origin of the instability, in the absence of an evident granular surface tension, has led to individuate grain-grain attractive interactions coming from a combination of van der Waals and capillary bridges between surface asperities. The role of granular temperature and the possibility of a dynamical surface tension induced by inelasticity has been ruled out.

The effect of gas entrainment is crucial also for the developement of the so-called *granular jets* . This beautiful phenomenon occurs when a solid sphere impacts on a deep layer of granular medium [85]. The impact produces a cylindrical cavity in the material, which subsequently collapses. The axisymmetrical collapse towards the center of the cavity generates a pressure spike which drives up the granular material in a narrow and high jet along the axis of symmetry, see Fig. 1.1 [75].

1.2.2 Shear

Classical studies of granular shear rheology are carried out in the common Couette geometry.

Even if there were earlier important experimental studies on the flow properties of granular materials (mainly initiated by Hagen [33] and Reynolds [72]), the modern pioneering work on the constitutive behavior of rapid granular flows was Bagnold's study [2] of wax spheres, suspended in a glycerin-water-alcohol mixture and sheared in a coaxial cylinder rheometer (Couette experiment). His main finding is a constitutive relation between internal stresses $\mathscr{T}_{ij}$ (where i and j denote Cartesian components, e.g. x, y and z) and shear rate γ:

$$\mathscr{T}_{ij} = \rho_p \sigma^2 \gamma^2 \mathscr{G}_{ij}(\phi) \tag{1.1}$$

Fig. 1.1 Formation of a
granular "jet": the two
sequences (**a**)-(**b**)-(**c**) and
(**d**)-(**e**)-(**f**) differ in the air
pressure. Reprinted by
permission from Macmillan
Publishers Ltd: Nature
Physics 1, page 164 [75],
copyright 2005

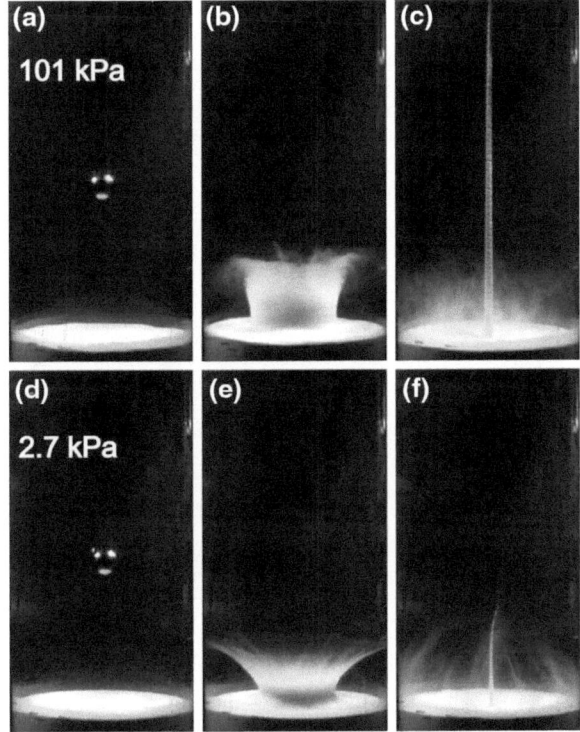

with ρ_p the particle density, σ the particle radius and $\mathscr{G}_{ij}$ a tensor-valued function of
the solid fraction ϕ. This relation has been confirmed in shear-cell experiments with
both wet or dry mixtures.

Bagnold measured not only shear stresses (i.e. transversal components, say $i \neq j$
in $\mathscr{T}_{ij}$), but also normal stress ($i = j$), that is the analogous of pressure in gas
kinetics: he referred to them as "dispersive stresses" as they tend to cause dilation of
the material.

Other experiments have focused on different phenomena observed in the Couette
rheometer.

- Fluctuations of stresses: already in [82] large fluctuations of internal (normal)
  stresses were observed; in [34] (see Fig. 1.2) a two-dimensional Couette experiment demonstrated that the mean internal stress follows a continuous transition
  when the packing fraction of the granular material changes and passes through a
  critical value $\phi_c = 0.776$: when the packing fraction is above the critical threshold
  the material shows strong fluctuations of internal stress, while under the threshold
  the stresses are averagely zero and the system is highly compressible.
- Microstructure in 3d: the bulk microstructure was studied in the dense shearing
  regime in a 3D Couette rheometer, using non-invasive imaging by X-Ray microto-
  mographyin [61]; it appears that the velocity parallel to the shear direction decays

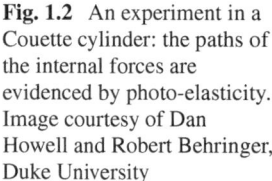

Fig. 1.2 An experiment in a
Couette cylinder: the paths of
the internal forces are
evidenced by photo-elasticity.
Image courtesy of Dan
Howell and Robert Behringer,
Duke University

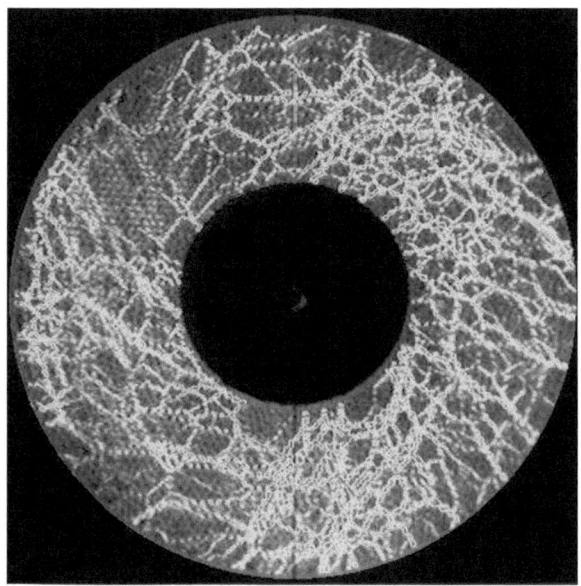

more rapidly than linear (from exponential to Gaussian-like decay, depending upon
the regularity of the grains). A similar strong decay of the flow with the distance
from the moving wall was observed in many experiments, for example in [51].

- Diluted (air-fluidized) shear: Couette experiments may be performed with a flow
  of air coming from the bottom of the cylinder, in order to fluidize the material and
  obtaining smoother profiles of the shear stress $\mathscr{T}(y)$ [53]. It is seen that the RMS
  fluctuations of velocity and the shear forces are related as $\mathscr{T}^{1/2}(y) \sim \gamma(y)^{\alpha}$ with
  $\alpha = \simeq 0.4$.

- Size segregation: convection patterns and size segregation are common in Couette
  flows, for instance in [40] the authors check the effect of interstitial fluids, finding
  it irrelevant.

- Planetary rings: planetary rings (those of Saturn for example) have been sometime
  studied in the framework of granular rheology, whereas the "geometry" of the
  planetary experiment is similar to a Couette cell (grains are circularly sheared
  because the angular velocity depends upon the distance from the planet). A review
  of these studies can be found in [6].

Another way to induce shear is making the granular flow along an inclined channel. In this kind of experiments the whole material is accelerated by gravity, but the
friction with the plane induce shearing, so that measurements similar to the ones
performed in Couette cells can be performed [15, 73, 79]. Interest has focused on
constitutive relations, as before, but also on the profiles of the hydrodynamic fields,
mainly flow velocity and solid fraction: computer simulations (see for example [8]
and for an exhaustive review the classical work of Campbell [7]) have allowed the
measurement of the temperature field: this has confirmed the picture of a fluid-like

behavior, explaining the reduction of density (solid fraction) near the bottom by means of an increase of granular temperature, due to the shear work. In this framework the scheme representing the "mechanical energy path" sketched by Campbell in his review on rapid granular flow [7] is enlightening. The external driving force (i.e. gravity) induces mean motion (kinetic energy) which consequently generates friction with boundaries, that is shear work (granular temperature). The randomization represented by the granular temperature induces collisions among the grains, which are dissipative. Moreover, granular temperature generates internal (transversal as well as normal) stresses.

Another configuration of granular flow under the force of gravity is the simple hopper geometry (a hopper is a funnel-shaped container in which materials, such as grain or coal, are stored in readiness for dispensation). The bottom of a hopper is opened and the grains start to pour out. As already discussed, the pressure (and therefore the flow rate) does not depend upon the height of the column of material. However, the flux of grains leaving the container produces complex flow regions inside the container. Four regions of density and velocity can be identified, most notably a tongue of dense motion just above the aperture and an area of no grain motion below a cone extending upwards from the opening (a similar effect can be observed in a silo, see Fig. 1.3). For large opening angles, density waves propagate upward from above the aperture against the direction of particle flow, but downwards for small angles [4] . The flow can even stop due to "clogging", i.e. the grains can form big arches above the aperture and sustain the entire weight of the column.

Other experiments have been performed on granular flows along inclined planes or chutes, evidencing several phenomena.

Fig. 1.3 Image of convergent flow of grains in a silo draining through an orifice. The silo was initially filled in horizontal layers with glass beads with two different colors but otherwise similar properties. Image courtesy: Azadeh Samadani and Arshad Kudrolli

- Validations of kinetic theory: in [10] an experiment of grain flow along an inclined channel was used to study the stationary profiles of velocity, solid fraction and granular temperature. The authors verify that there is a limited range of inclinations of the channel that allow for a stationary flow. Moreover they have probed the validity of the kinetic theories developed in the previous years [36, 37, 55, 56, 81], based on the assumption of slight perturbation to the Maxwellian equilibrium. The profiles of hydrodynamic fields show two different regions: a collisional region (higher density) where the transport is mainly due to collisions, and a ballistic region (on the upper free surface) where the grains fly almost ballistically.
- Size segregation in silo filling or emptying: the authors of [78] have studied the phenomena of size segregation in a quasi-two dimensional silo emptying out of an orifice. They have also studied the effects of interstitial fluids [77].
- Size segregation in rotating drums: another typical experiment, inspired to many industrial situations, is the tumbling mixer, or rotating drum, i.e. a container with some shape that rotate around a fixed axis, usually used to mix different kind of granular materials (typically powders in the pharmaceutical, chemical, ceramic, metallurgical and construction industry). Depending on the geometry of the mixer, the shapes of the grains, the parameters of the dynamics and so on, the grains can mix or separate. A very large literature exists on this phenomena (see the review in [66]). Usually, segregation is strictly tied to convection: there is a shallow flowing layer on the surface of the material inside the rotating drum, the grains at the end of it are transported into the bulk and follow a convective path so that they emerge again in another point of the surface. Segregation happens in many different ways: segregated bands appear and slowly enlarge (like in a coarsening model), segregation can emerge in different directions, e.g. parallel to the rotation axis as well as transversal to it.
- Shear stress fluctuations and frictional (stick-slip) properties of a granular medium has been studied in a ring-shaped cell with a rotating cover, finding agreement with a simple Brownian model similar to those underlying the Barkhausen effect in ferromagnets [3].

1.2.3 Shakers

Many interesting observations can be done when the granular medium is subject to vertical vibration, usually under the effect of gravity. As already mentioned the effect of slow vibration of a container filled of grains induces a very slow compaction of the material. When the amplitude of vibration is strong enough, i.e. when

$$\Gamma = \frac{a_{\max}}{g} > 1 \tag{1.2}$$

(where a_{max} is the maximum acceleration of the vibrating plate, e.g. $a_{max} = A\omega^2$ if the plate is harmonically vibrating with A amplitude and ω frequency) the granular shows several phenomena.

- Convection and segregation: A large literature [23] exists on the convection and segregation phenomena observed in granular media contained in a shaken box. Faraday [24] was perhaps the first to observe such a phenomenon. The geometry of the container can change dramatically the quality of the convection (e.g. in a cylinder may happen that the grains near the walls move downwards and the ones in the bulk move upwards, while inside an inverted cone the convection occurs in the opposite direction). Usually, the larger grains (independent of their density) tend to move upwards, so that the material segregate (see for example [16, 41–43, 48]).
- Pattern formation in surface waves: another problem that has been extensively studied in recent years is the formation of patterns on the surface of vibrated layers of grains. Depending on the whole set of parameters (amplitude and frequency of the vibration, shapes and sizes of the grains, size of the container, depth of the bed and so on) different qualities of standing waves can be observed, leading to unexpected and fascinating textures [58–60, 87] (see Figs. 1.4 and 1.5).
- Clustering: in [44, 46] the formation of clusters was studied, measuring the density distribution in an experiment consisting of steel balls rolling on a smooth surface which could or could not be inclined with a vibrating side. The experiment took into account a monolayer (not completely covered) of grains, in order to study a true $2d$ setup. In both cases (inclined or horizontal), at high enough global densities, the distribution of density (going from Poissonian to exponential) indicates strong clustering. The formation of high density clusters has also been studied in a vibrated cylindrical piston [20–22]. A transition has been observed with the increasing number of particles in the cylinder, from a gas-like behavior to a collective solid-like behavior. Such a transition has been also observed in the framework of fluidized beds [64], i.e. vertically shaken granular monolayers: the authors have observed

Fig. 1.4 Different surface patterns obtained by vertical vibration of granular layers. © 1996 Paul B. Umbanhowar. All rights reserved

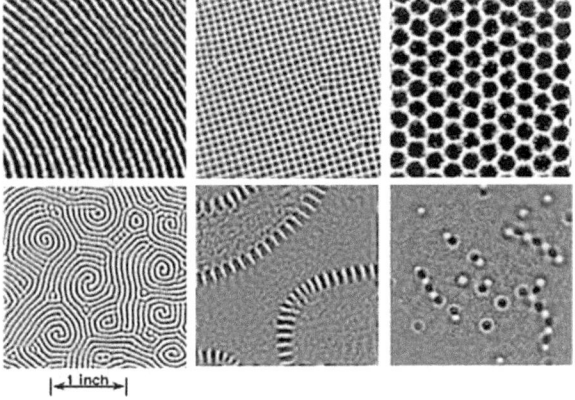

|◄1 inch►|

a transition (with reducing the vibration amplitude) from a gas-like motion to a coexistence of a crystallized state (a pack of particles arranged in an ordered way) and a gas.

- The Leidenfrost effect is a particular stationary configuration, obtained under gravity and vertical shaking, consisting in a granular "drop" at high packing fraction floating above a more dilute granular gas which is in direct contact with the vibrating boundary [19]. In liquids this effect is known since the 18th century and is encountered when a liquid drop is let in contact with a surface at a temperature much higher than its boiling point. In cooking, it is common to observe such an effect when sprinkling water droplets on the hot surface of a pan: if the pan is hot enough, drops skitter across the surface. The drop takes a long time to evaporate because of a thin vapour layer that isolate them from direct contact with the pan.
- A systematic study in a vertically vibrated quasi-two dimensional container with length and height much larger then depth, allowed to trace a quite robust phase diagram [18]. The experiment showed a wide variety of phenomena: bouncing bed, undulations, granular Leidenfrost effect, convection rolls, and granular gas. These phenomena and the transitions among them are characterized by a few main control parameters: the shaking maximum acceleration Γ, the number of bead layers F,

Fig. 1.6 The phase diagram in the experiment with a quasi-2d vertically vibrated container. Reproduced with permission from Eshuis et al. [18]. © 2007, AIP Publishing LLC

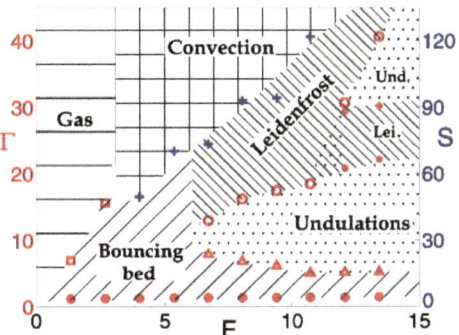

the inelasticity parameter $1 - r^2$ where r is the restitution coefficient, and the aspect ratio, i.e. the ratio between the length of the container and the height of the granular media at rest. The authors studied in particular the effect of Γ and F, obtaining a phase diagram of the kind in Fig. 1.6. The diagram slightly changes when other parameters are modified.

- In the context of random lasers, shaken granular lasers have appeared recently [27, 28]. The general idea of random lasers consists in pumping light through a scattering/amplifying random medium. Emitted light display a spectrum with random peaks which depend on many parameters of the scattering system. In this framework, a shaken granular laser is obtained by vibrating a cell which contains glass or steel spheres (1 mm diameter) dispersed in a "gain medium", i.e. a light-amplifying fluid such as a rhodamine solution. The added value of shaking is in a direct control of statistical properties of emitted spectra: different choices of shaking parameters lead to different stationary regimes with more dilute or dense granular assemblies.

- Validations of kinetic theory: a part of the experimental effort [54, 90–92] has also devoted to the study of hydrodynamic and kinetics fields (i.e. packing fraction profiles, granular temperature profiles, self-diffusion, velocity statistics) in vertically vibrated boxes (or vertical slices, that is 2d setups). The interest has also focused on the difficulties of imposing boundary conditions to the existing kinetics model, due to the existence of non-hydrodynamic boundary layers. This has also led to the formulation of hypothesis of scaling for the granular temperature as a function of the amplitude of vibration [47, 83]. For more recent experiments see [93].

- Non-Gaussian velocity distributions: after the evidences found in the numerical study of granular rapid dynamics, the question of the true form of the velocity distributions has arisen and has induced many new experiments in order to give an answer to it. In [45] the distributions of velocities were studied along an inclined plane with varying angles of inclination, obtaining non-Gaussian statistics with enhanced high energy tails; it was seen that increasing the angle of inclination the distributions tends toward the Maxwellian. The experiment reported in [64, 65] with a horizontal granular monolayer subject to a vertical

Fig. 1.7 Probability distributions of horizontal velocities of grains in a vertically shaken granular monolayer. Reprinted with permission from Olafsen and Urbach [64]. Copyright 1998 by the American Physical Society

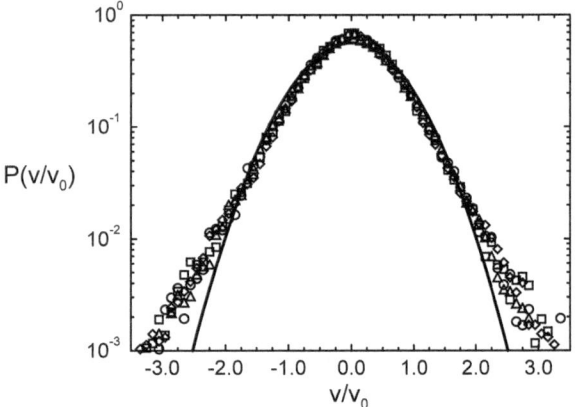

vibration (and measuring horizontal velocities) has proven that, in the presence of clustering, the distributions are non-Gaussian, showing nearly exponential tails (see Fig. 1.7). A different experiment [52] on a similar monolayer with vertical vibration verified that both the theoretical predictions of [88] on the high energy tails for cooling and driven granular gases are correct, measuring exponential tails for the former and $\exp(-v^{3/2})$ for the latter. More recently [74], the velocity fluctuations in a vertically vibrated vertical monolayer of grains have been measured, obtaining again a velocity distribution with $\exp(-v^{3/2})$ tails.

- Non-equilibrium behavior: a few experiments have been devoted to the study of non-equilibrium granular properties. In particular two experiments have verified the breakdown of energy equipartition [25] and have measured the fluctuations of internal energy flow [26]: in the last experiment the authors claimed a verification of the Gallavotti-Cohen Fluctuation theorem [29], but successive theoretical work has proven that it was not necessarily the case [68]; more recently, the exchanges of energy between a granular gas and a harmonic oscillator, in the stationary state [62] or during periodic cycles between different forcings [57], have been measured.

- Velocity correlations: experiments similar to the setup of [45] has revealed strong correlations between velocity particles [5]. More recently, velocity structure factors, in good agreement with fluctuating granular hydrodynamics, have been measured at average packing fraction (30–40 %) in a monolayer of spheres moving on a vertically vibrating horizontal rough plate [32, 67]. The measured velocity correlations are characterized by a correlation length which increases with the packing fraction.

- Linear response has been studied in similar experiments where a Brownian rotator is suspended in a granular gas and is excited with a small torque. The first experiment [12] probed a very dense system over long time-scales, observing the validity of a fluctuation-dissipation relation similar to equilibrium, with an effective temperature. More recently, a similar experiment has been carried out in dilute and averagely dense configurations [30]: it has been possible to put in evidence the entanglement between fast and slow time-scales, which induce non-equilibrium correlations growing with the density and a consequent breakdown of the Einstein relation, which is equivalent to the fluctuation-dissipation relation at equilibrium.

- A series of experiments under gravity and vertical vibration, in container separated into communicated chambers (compartments), have demonstrated the tendency of granular fluids to violate many entropic trends of fluids at equilibrium [50]. Typical examples are phase separation (dense versus dilute) between different compartments, and spontaneous segregation of mixtures into different compartments. These scenario have been often assimilated to the realization of a "Maxwell Demon" experiment.

- Conceptually similar to the Maxwell Demon phenomenon illustrated above, the realization of a granular ratchet more strikingly illustrates the rectification of unbiased fluctuations under non-equilibrium conditions. Models have been proposed since [11], but the first experimental realization has been obtained in [17]. More recently an experiment in fair quantitative agreement with kinetic theory have been carried out, with surprising results about the crucial role of Coulomb friction [31].

1.3 Granular Versus Active Fluids

Younger than granular materials, the study of active fluids reveals several analogies in the observed phenomena and in the methods of investigation. Under the name of active fluid or active matter one includes a large class of systems, mostly in the realm of living organisms, involving a large number of "self-propelled" constituents [70]. Examples include biofilaments and molecular motors in vitro or in vivo [38], collections of motile microorganisms such as alga blooms or biofilms, bird flocks and fish schools [86], and chemical or mechanical imitations.

The fundamental ingredient in active fluids is the propulsion: the elementary constituents, or "particles", displace themselves by walking, crawling, swimming, flying, etc. An active particle is able to transform energy from a reservoir into directed motion. The most famous example is, perhaps, that of molecular motors inside cells, where the propulsion is obtained through chemical reactions involving the hydrolysis of ATP. Organisms at all lengthscales (from bacteria to birds and fishes) display mechanisms of propulsion of great variety, complexity and beauty. For active fluids, however, the details of the propelling device are not the main object of study. Researchers are instead interested in the collective properties of *many* particles under given boundary conditions [89]. The presence or absence of a solvent fluid determines the kind of interaction among the particles, which can be dissipative or conservative, at contact or long-range. Self-propelled particles dispersed in a fluid, usually provided with a well-defined polarity, are also called "swimmers". In general they are divided in two categories: pushers and pullers, depending on the origin of the movement, i.e. from the front of from the rear (this has consequences on the kind of flow generated in the surrounding fluid and therefore on the interaction with other particles).

A collection of self-propelled particles is intrinsically out-of-equilibrium. Each particle *injects* kinetic energy into the system, as the result of conversion from an external reservoir. Viscous dissipation through the solvent, or—more rarely—non-conservative interactions with surrounding particles, balances the injection and may determine a statistically steady state. Such a balance of energy currents is reminiscent of the *energy cycle* in shaken granular fluids: grains take energy from hitting the moving walls of a shaken container, and dissipate it through collisions. A closer look suggests that a granular fluid is often an *anti-active fluid*: indeed the smallest length-scale (the grain or the particle) is responsible for dissipation in granular fluid, while it is the source of energy in active fluids. However there are examples where the analogy is even more fitting (see for instance the thermostat model discussed in Sect. 2.3.5).

The most common phenomena observed in active fluids is one or more ordering transitions. These transitions, typically toward a polar (i.e. ferromagnetic) or nematic order, occur when parameters such as propulsion velocity or particle density are changed. The elonged shape which allows to define the "direction" of a self-propelled particle, is perhaps the main difference with respect to the large majority of granular systems. The reader, however, will notice that the shear instability in cooling granular materials, where clusters of grains appear to move as "swarms", see Sect. 3.3.1, is not dissimilar from collective behavior in many models of active particles for bacterial suspensions and colonies [86].

The analogy between active and granular fluids have been pursued in some experiments. The granular particles were shaped in such a way that isotropy was broken, and put above a vertically vibrated horizontal plate. The anisotropy, together with subtle frictional mechanisms and vertical vibration, produced a self-propulsion effect. Several properties, including collective "swarming"-like effects, have been demonstrated [14].

The strongest analogy with granular materials is, however, in the methods of investigation. Hydrodynamics, which is extensively discussed in Chap. 3 of this book, is used to describe many collective phenomena in granular fluids as well as in active fluids [71]. The working principle is the same: a few slowly evolving observables are identified and transport equations for these observables are built, based upon more or less rigorous kinetic theories. Symmetry arguments are usually sufficient to determine the basic structure of these equations, while more refined (microscopic) calculations are necessary to assign values to the transport coefficients. Even the qualitative form of hydrodynamic equations, with approximate orders of magnitude for the transport coefficients, are enough to catch the stable states and the transitions between them.

Active hydrodynamics faces risks or problems similar to those encountered in granular hydrodynamics: the number of microscopic constituents are not huge, the separation of scale between macroscopic and microscopic lengths or times is not always clear and guaranteed. In many situations huge fluctuations are observed [69] and a stochastic treatment of the macroscopic equations becomes hardly avoidable. Fluctuating hydrodynamics for active fluids is still in its infancy [84], while a few important steps have been carried out for granular systems.

References

1. Andreotti, B., Forterre, Y., Pouliquen, O.: Granular Media. Cambridge University Press, Cambridge (2013)
2. Bagnold, R.A.: Experiments on a gravity-free dispersion of large solid spheres in a Newtonian fluid under shear. Proc. Royal Soc. Lond. **225**, 49 (1954)
3. Baldassarri, A., Dalton, F., Petri, A., Zapperi, S., Pontuale, G., Pietronero, L.: Brownian forces in sheared granular matter. Phys. Rev. Lett. **96**, 118002 (2006)
4. Baxter, G.W., Behringer, R.P., Fagert, T., Johnson, G.A.: Pattern formation in flowing sand. Phys. Rev. Lett. **62**, 2825 (1989)
5. Blair, D.L., Kudrolli, A.: Velocity correlations in dense granular gases. Phys. Rev. E **64**, 050301 (2001)
6. Brahic, A: Dynamical evolution of viscous discs. Astrophysical applications to the formation of planetary systems and to the confinement of planetary rings and arcs. In: Pöschel, T., Luding S. (eds.) Granular Gases, volume 564 of Lectures Notes in Physics. Springer, Berlin (2001)
7. Campbell, C.S.: Rapid granular flows. Ann. Rev. Fluid Mech. **22**, 57 (1990)
8. Campbell, C.S., Brennen, C.E.: Computer simulation of granular shear flows. J. Fluid. Mech. **151**, 167 (1985)
9. Cheng, X., Xu, L., Patterson, A., Jaeger, H.M., Nagel, S.R.: Towards the zero-surface-tension limit in granular fingering instability. Nat. Phys. **4**, 234 (2008)
10. Chevoir, F., Azanza, E., Moucheront, P.: Experimental study of collisional granular flows down an inclined plane. J. Fluid Mech. **400**, 199 (1998)
11. Costantini, G., Puglisi, A., Marini Bettolo Marconi, U.: A granular brownian ratchet model. Phys. Rev. E **75**, 061124 (2007)
12. D'Anna, G., Mayor, P., Barrat, A., Loreto, V., Nori, F.: Observing brownian motion in vibration-fluidized granular matter. Nature **424**, 909 (2003)
13. de Gennes, P.G.: Granular matter: a tentative view. Rev. Mod. Phys. **71**, S374 (1999)
14. Deseigne, J., Dauchot, O., Chaté, H.: Collective motion of vibrated polar disks. Phys. Rev. Lett. **105**, 098001 (2010)
15. Drake, T.G.: Structural features in granular flows. J. Geophys. Res. **95**, 8681 (1990)
16. Ehrichs, E.E., Jaeger, H.M., Karczmar, G.S., Knight, J.B., Kuperman, V.Yu., Nagel, S.R.: Granular convection observed by magnetic resonance imaging. Science **267**, 1632 (1995)
17. Eshuis, P., van der Weele, K., Lohse, D., van der Meer, D.: Experimental realization of a rotational ratchet in a granular gas. Phys. Rev. Lett. **104**, 248001 (2010)
18. Eshuis, P., van der Weele, K., van der Meer, D., Bos, R, Lohse, D.: Phase diagram of vertically shaken granular matter. Phys. Fluids **19**, 123301 (2007)
19. Eshuis, P., van der Weele, K., van der Meer, D., Lohse, D.: Granular Leidenfrost effect: experiment and theory of floating particle clusters. Phys. Rev. Lett. **95**, 258001 (2005)
20. Falcon, E., Fauve, S., Laroche, C.: Cluster formation, pressure and density measurements in a granular medium fluidized by vibrations. Eur. Phys. J. B. **9**, 183 (1999)
21. Falcon, E., Fauve, S., Laroche, C.: An experimental study of a granular gas fluidized by vibrations. In Pöschel, T., Luding, S. (eds.) Granular Gases, volume 564 of Lectures Notes in Physics. Springer, Berlin (2001)
22. Falcon, E., Wunenburger, R., Evesque, P., Fauve, S., Chabot, C., Garrabos, Y., Beysens, D.: Cluster formation in a granular medium fluidized by vibrations in low gravity. Phys. Rev. Lett. **83**, 440 (1999)

23. Fan, L.T., Chen, Y.-M., Lai, F.S.: Recent developments in solids mixing. Powder Technol. **61**, 255 (1990)
24. Faraday, M.: On a peculiar class of acoustical figures; and on certain forms assumed by groups of particles upon vibrating elastic surfaces. Philos. Trans. R. Soc. Lond. **52**, 299 (1831)
25. Feitosa, K., Menon, N.: Breakdown of energy equipartition in a 2d binary vibrated granular gas. Phys. Rev. Lett. **88**, 198301 (2002)
26. Feitosa, K., Menon, N.: Fluidized granular medium as an instance of the fluctuation theorem. Phys. Rev. Lett. **92**, 164301 (2004)
27. Folli, V., Ghofraniha, N., Puglisi, A., Leuzzi, L., Conti, C.: Time-resolved dynamics of granular matter by random laser emission. Sci. Rep. **3**, 2251 (2013)
28. Folli, V., Puglisi, A., Leuzzi, L., Conti, C.: Shaken granular lasers. Phys. Rev. Lett. **108**, 248002 (2012)
29. Gallavotti, G., Cohen, E.G.D.: Dynamical ensembles in stationary states. J. Stat. Phys. **80**, 931 (1995)
30. Gnoli, A., Puglisi, A., Sarracino, A., Vulpiani, A.: Nonequilibrium brownian motion beyond the effective temperature. Plos One **9**, e93720 (2014)
31. Gnoli, A., Petri, A., Dalton, F., Gradenigo, G., Pontuale, G., Sarracino, A., Puglisi, A.: Brownian ratchet in a thermal bath driven by coulomb friction. Phys. Rev. Lett. **110**, 120601 (2013)
32. Gradenigo, G., Sarracino, A., Villamaina, D., Puglisi, A.: Non-equilibrium length in granular fluids: from experiment to fluctuating hydrodynamics. Europhys. Lett. **96**, 14004 (2011)
33. Hagen, G.H.L.: Über den Druck und die Bewegung des trockenen Sandes. Monatsberichte der königlich, preußischen Akademie der Wissenschaften zu Berlin, p 35, 19 Jan 1852
34. Howell, D., Behringer, R.P., Veje, C.: Stress fluctuations in a 2d granular Couette experiment: a continuous transition. Phys. Rev. Lett. **82**, 5241 (1999)
35. Jaeger, H.M., Nagel, S.R., Behringer, R.P.: Granular solids, liquids, and gases. Rev. Mod. Phys. **68**, 1259 (1996)
36. Jenkins, J.T., Richman, M.W.: Kinetic theory for plane shear flows of a dense gas of identical, rough, inelastic, circular disks. Phys. Fluids **28**, 3485 (1985)
37. Jenkins, J.T., Savage, S.B.: A theory for the rapid flow of identical, smooth, nearly elastic, spherical particles. J. Fluid Mech. **130**, 187 (1983)
38. Jlicher, F., Kruse, K., Prost, J., Joanny, J.-F.: Active behavior of the cytoskeleton. Phys. Rep. **449**, 3 (2007)
39. Keys, A.S., Abate, A.R., Glotzer, S.C., Durian, D.J.: Measurement of growing dynamical length scales and prediction of the jamming transition in a granular material. Nat. Phys. **3**, 260 (2007)
40. Khosropour, R., Zirinsky, J., Pak, H.K., Behringer, R.P.: Convection and size segregation in a couette flow of granular material. Phys. Rev. E **56**, 4467 (1997)
41. Knight, J.B.: External boundaries and internal shear bands in granular conveciton. Phys. Rev. E **55**, 6016 (1997)
42. Knight, J.B., Ehrichs, E.E., Kuperman, VYu., Flint, J.K., Jaeger, H.M., Nagel, S.R.: Experimental study of granular convection. Phys. Rev. E **54**, 5726 (1996)
43. Knight, J.B., Jaeger, H.M., Nagel, S.R.: Vibration-induced size separation in granular media: the convection connection. Phys. Rev. Lett. **70**, 3728 (1993)
44. Kudrolli, A., Gollub, J.P.: Studies of cluster formation due to collisions in granular material. In: Powders & Grains 97, p 535, Rotterdam, Balkema (1997)
45. Kudrolli, A., Henry, J.: Non-gaussian velocity distributions in excited granular matter in the absence of clustering. Phys. Rev. E **62**, R1489 (2000)
46. Kudrolli, A., Wolpert, M., Gollub, J.P.: Cluster formation due to collisions in granular material. Phys. Rev. Lett. **78**, 1383 (1997)
47. Kumaran, V.: Temperature of a granular material "fluidized" by external vibrations. Phys. Rev. E **57**, 5660 (1998)
48. Kuperman, VYu., Ehrichs, E.E., Jaeger, H.M., Karczmar, G.S.: A new technique for differentiating between diffusion and flow in granular media using magnetic resonance imaging. Rev. Sci. Instrum. **66**, 4350 (1995)

49. Liu, A.-J., Nagel, S.R.: Nonlinear dynamics: jamming is not just cool any more. Nature **396**, 21 (1998)
50. Lohse, D., Rauhè, R.: Creating a dry variety of quicksand. Nature **432**, 689 (2004)
51. Losert, W., Bocquet, L., Lubensky, T.C., Gollub, J.P.: Particle dynamics in sheared granular matter. Phys. Rev. Lett. **85**, 1428 (2000)
52. Losert, W., Cooper, D.G.W., Delour, J., Kudrolli, A., Gollub, J.P.: Velocity statistics in vibrated granular media. Chaos **9**, 682 (1999)
53. Lubensky, T.C. Losert, W., Bocquet, L. Gollub, J.P.: Particle dynamics in sheared granular matter. Phys. Rev. Lett. **85**, 1428 (2000)
54. Luding, S., Clément, E., Blumen, A., Rajchenbach, J., Duran, J.: Studies of columns of beads under external vibrations. Phys. Rev. E **49**, 1634 (1994)
55. Lun, C.K.K.: Kinetic theory for granular flow of dense, slightly inelastic, slightly rough spheres. J. Fluid Mech. **233**, 539 (1991)
56. Lun, C.K.K., Savage, S.B., Jeffrey, D.J., Chepurniy, N.: Kinetic theories for granular flow: inelastic particles in couette flow and slightly inelastic particles in a general flowfield. J. Fluid Mech. **140**, 223 (1984)
57. Mounier, A., Naert, A.: The Hatano-Sasa equality: transitions between steady states in a granular gas. Europhys. Lett. **100**, 30002 (2012)
58. Melo, F., Umbanhowar, P.B., Swinney, H.L.: Transition to parametric wave patterns in a vertically oscillated granular layer. Phys. Rev. Lett. **72**, 172 (1994)
59. Melo, F., Umbanhowar, P.B., Swinney, H.L.: Hexagons, kinks, and disorder in oscillated granular layers. Phys. Rev. Lett. **75**, 3838 (1995)
60. Metcalf, T.H., Knight, J.B., Jaeger, H.M.: Standing wave patterns in shallow beds of vibrated granular material. Physica A **236**, 202 (1997)
61. Mueth, Daniel M., Debregeas, Georges F., Karczmar, Greg S., Eng, Peter J., Nagel, Sidney R., Jaeger, Heinrich M.: Signatures of granular microstructure in dense shear flows. Nature **406**, 385 (2000)
62. Naert, A.: Experimental study of work exchange with a granular gas: the viewpoint of the fluctuation theorem. Europhys. Lett. **97**, 20010 (2012)
63. Ogawa, S.: Multitemperature theory of granular materials. In: Cowin, S.C., Satake, M. (eds) Proceedings of US-Japan Symposium on Continuum Mechanics and Statistical Approaches to the Mechanics of Granular Media, p. 208. Fukyu-kai, Gakujutsu Bunken (1978)
64. Olafsen, J.S., Urbach, J.S.: Clustering, order and collapse in a driven granular monolayer. Phys. Rev. Lett. **81**, 4369 (1998)
65. Olafsen, J.S., Urbach, J.S.: Velocity distributions and density fluctuations in a 2d granular gas. Phys. Rev. E **60**, R2468 (1999)
66. Ottino, J.M., Khakhar, D.V.: Mixing and segregation of granular materials. Ann. Rev. Fluid Mech. **32**, 55 (2000)
67. Puglisi, A., Gnoli, A., Gradenigo, G., Sarracino, A., Villamaina, D.: Structure factors in granular experiments with homogeneous fluidization. J. Chem. Phys. **136**, 014704 (2012)
68. Puglisi, A., Visco, P., Barrat, A., Trizac, E., van Wijland, F.: Fluctuations of internal energy flow in a vibrated granular gas. Phys. Rev. Lett. **95**, 110202 (2005)
69. Ramaswamy, S., Simha, R.A., Toner, J.: Active nematics on a substrate: Giant number fluctuations and long-time tails. Europhys. Lett. **62**, 196 (2003)
70. Ramaswamy, S.: The mechanics and statistics of active matter. Annu. Rev. Condens. Matter Phys. **1**, 323 (2010)
71. Ramaswamy, S. Liverpool, T.B., Prost, J., Rao, M., Marchetti, M.C., Joanny, J.F., Aditi Simha, R.: Hydrodynamics of soft active matter. Rev. Mod. Phys. **85**, 1143 (2013)
72. Reynolds, O.: On the dilatancy of media composed of rigid particles in contact. Philos. Mag. Ser. 5 **50–20**, 469 (1885)
73. Ridgway, K., Rupp, R.: Flow of granular material down chutes. Chem. Proc. Eng. **51**, 82 (1970)
74. Rouyer, F., Menon, N.: Velocity fluctuations in a homogeneous 2d granular gas in steady state. Phys. Rev. Lett. **85**, 3676 (2000)

75. Royer, J.R., Corwin, E.I., Flior, A., Cordero, M.-L., Rivers, M.L., Eng, P.J., Jaeger, H.M.: Formation of granular jets observed by high-speed X-ray radiography. Nat. Phys. **1**, 164 (2005)
76. Royer, J.R., Evans, D.J., Oyarte, L., Guo, Q., Kapit, E., Moöbius, M.E., Waitukaitis, S.R., Jaeger, H.M.: High-speed tracking of rupture and clustering in freely falling granular streams. Nature **459**, 1110 (2009)
77. Samadani, A., Kudrolli, A.: Segregation transitions in wet granular matter. Phys Rev. Lett. **85**, 5102 (2000)
78. Samadani, A., Pradhan, A., Kudrolli, A.: Size segregation of granular matter in silo discharges. Phys. Rev. E **60**, 7203 (1999)
79. Savage, S.B.: Gravity flow of cohesionless granular materials in chutes and channels. J. Fluid Mech. **92**, 53 (1979)
80. Savage, S.B.: The mechanics of rapid granular flows. Adv. Appl. Mech. **24**, 289 (1984)
81. Savage, S.B., Jeffrey, D.J.: The stress tensor in a granular flow at high shear rates. J. Fluid. Mech. **110**, 255 (1981)
82. Savage, S.B., Sayed, M.: Stresses developed by dry cohesionless granular materials sheared in an annular shear cell. J. Fluid Mech. **142**, 391 (1984)
83. Sunthar, P., Kumaran, V.: Temperature scaling in a dense vibrofluidized granular material. Phys. Rev. E **60**, 1951 (1999)
84. Solon, A.P., Tailleur, J.: Revisiting the flocking transition using active spins. Phys. Rev. Lett. **111**, 078101 (2013)
85. Thoroddsen, S.T., Shen, A.Q.: Granular jets. Phys. Fluids **13**, 4 (2001)
86. Toner, J., Tu, Y., Ramaswamy, S.: Hydrodynamics and phases of flocks. Ann. Phys. **318**, 170 (2005)
87. Umbanhowar, P.B., Melo, F., Swinney, H.L.: Localized excitations in a vertically vibrated granular layer. Nature **382**, 793 (1996)
88. van Noije, T.P.C., Ernst, M.H.: Velocity distributions in homogeneous granular fluids: the free and the heated case. Granular Matter **1**, 57 (1998)
89. Vicsek, T., Czirk, A., Ben-Jacob, E., Cohen, I., Shochet, O.: Novel type of phase transition in a system of self-driven particles. Phys. Rev. Lett. **75**, 1226 (1995)
90. Warr, S., Huntley, J.M.: Energy input and scaling laws for a single particle vibrating in one dimension. Phys. Rev. E **52**, 5596 (1995)
91. Warr, S., Huntley, J.M., Jacques, G.T.H.: Fluidization of a two-dimensional granular system: experimental study and scaling behavior. Phys. Rev. E **52**, 5583 (1995)
92. Warr, S., Jacques, G.T.H., Huntley, J.M.: Tracking the translational and rotational motion of granular particles: use of high-speed photography and image processing. Powder Technol. **81**, 41 (1994)
93. Wildman, R.D., Huntley, J.M., Parkar, D.J.: Convection in highly fluidized three-dimensional granular beds. Phys. Rev. Lett. **86**, 3304 (2001)

Chapter 2
Boltzmann Equation: A Gas of Grains

Abstract A simple but realistic and rich model for fluidized granular media is the gas of inelastic hard spheres. In this chapter its statistical description is reviewed. A key role is played by the assumption of Molecular-Chaos and by the Boltzmann equation. A comparison with the case of elastic hard spheres is made, pointing out the analogies and the differences. The chapter is concluded with the discussion of the protocols used for energy injection.

2.1 Collisions

Let us consider two point-like particles with masses m_1 and m_2, coordinates $\mathbf{r}_1$ and $\mathbf{r}_2$ and velocities $\mathbf{v}_1$ and $\mathbf{v}_2$. One can introduce the center of mass vector $\mathbf{r}_c$:

$$\mathbf{r}_c = \frac{m_1 \mathbf{r}_1 + m_2 \mathbf{r}_2}{m_1 + m_2} \tag{2.1}$$

and the relative position vector:

$$\mathbf{r} = \mathbf{r}_1 - \mathbf{r}_2. \tag{2.2}$$

Their time derivatives are the velocity of the center of mass

$$\mathbf{v}_c = \frac{m_1 \mathbf{v}_1 + m_2 \mathbf{v}_2}{m_1 + m_2} \tag{2.3}$$

and the relative velocity

$$\mathbf{V}_{12} = \mathbf{v}_1 - \mathbf{v}_2. \tag{2.4}$$

The forces between these two particles depends only on their relative position and are of equal magnitude and pointing in opposite directions:

$$\mathbf{F}_{12}(\mathbf{r}) = -\mathbf{F}_{21}(\mathbf{r}). \tag{2.5}$$

This is equivalent to say that the center of mass does not accelerate, i.e.:

© The Author(s) 2015
A. Puglisi, *Transport and Fluctuations in Granular Fluids*,
SpringerBriefs in Physics, DOI 10.1007/978-3-319-10286-3_2

$$\frac{d^2\mathbf{r}_c}{dt^2} = 0 \qquad (2.6)$$

while the relative position obeys to the following equation of motion:

$$m^*\frac{d^2\mathbf{r}}{dt^2} = \mathbf{F}_{12}(\mathbf{r}) \qquad (2.7)$$

where

$$m^* = \left(\frac{1}{m_1} + \frac{1}{m_2}\right)^{-1} \qquad (2.8)$$

is the reduced mass of the system of two particles. If the "collision" is elastic an interaction potential can be introduced so that:

$$\mathbf{F}_{12} = -\frac{dU(r)}{dr}\hat{\mathbf{r}} \qquad (2.9)$$

where $\hat{\mathbf{r}}$ is the unit vector along the direction of the relative position of the two particles. The force vector lies in the same plane where the relative position vector and relative velocity vector lie. The evolution of the relative position r is the evolution of the position of a particle of mass m^* in a central potential $U(r)$. The angular momentum of the relative motion $\mathbf{L} = \mathbf{r} \times m^*\mathbf{V}_{12}$ is conserved. This means that the particle trajectory, during the collision, will be confined to this plane. Figure 2.1 sketches the typical binary scattering event when the interacting force is repulsive (monotonically decreasing potential), in the center of mass frame.

In the center of mass frame the elastic scattering has a very simple picture: the velocities of the particles are $\mathbf{v}_{1c} = \mathbf{V}_{12}m^*/m_1$ and $\mathbf{v}_{2c} = -\mathbf{V}_{12}m^*/m_2$. The elastic collision conserves the modulus of the relative velocity V_{12} and therefore also the moduli of the velocities of the particles in the center of mass frame. If one consider the collision event as a black box and observes the velocities of the particles "before" and "after" the interaction (i.e. asymptotically, when the interaction is negligible), then the velocity vectors are simply rotated of an angle χ called *angle of deflection*, which also represents the angle between asymptotic initial and final directions of the relative velocity. During the collision the total momentum is conserved (this holds for both elastic and inelastic collisions) but is redistributed between the two particles, i.e. the variation of the momentum of the particle 1 is $\delta(m_1\mathbf{v}_1) = m^*(\mathbf{V}'_{12} - \mathbf{V}_{12})$ where the prime indicates the post-collisional relative velocity. Obviously $\delta(m_1\mathbf{v}_1) = -\delta(m_2\mathbf{v}_2)$. Finally, one can calculate the components of the momentum transfer parallel and perpendicular to the relative velocity:

$$\delta(m_1\mathbf{v}_1)_\parallel = -m^*V_{12}(1 - \cos\chi) \qquad (2.10a)$$
$$\delta(m_1\mathbf{v}_1)_\perp = m^*V_{12}\sin\chi. \qquad (2.10b)$$

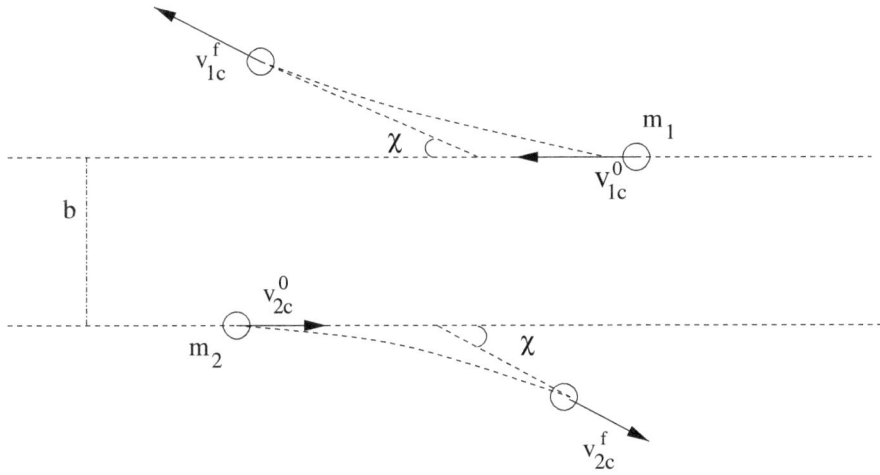

Fig. 2.1 The binary elastic scattering event in the center of mass frame, with a repulsive potential of interaction. The superscripts 0 and f denote initial and final velocities

To calculate the angle of deflection χ one needs the exact form of the interaction potential, the asymptotic initial relative velocity V_{12}^0 and the *impact parameter b* that is the minimal distance between the trajectories of the particles if there were no interaction between them:

$$\chi = \pi - 2 \int_{r_m}^{\infty} dr \frac{b}{r} \left[r^2 - b^2 - \frac{2r^2 U(r)}{m^*(V_{12}^0)^2} \right]^{-1/2} \tag{2.11}$$

where r_m is the closest distance effectively reached by the two particles. From Eq. (2.11) it is evident that the angle of deflection decreases as the initial relative velocity increases.

2.1.1 Elastic Smooth Hard Spheres

Two hard spheres in 3D (hard disks in 2D, hard rods in 1D) of diameters σ_1 and σ_2 interact by means of a discontinuous potential $U(r)$ of the form:

$$U(r) = 0 \quad (r > \sigma_{12}) \tag{2.12a}$$
$$U(r) = \infty \quad (r < \sigma_{12}) \tag{2.12b}$$

where $\sigma_{12} = (\sigma_1 + \sigma_2)/2 = r_m$ is the distance of the centers of the spheres at contact. The potential in Eqs. (2.12a, 2.12b) can be taken as a definition of hard spheres systems. In this case the deflection angle is given by

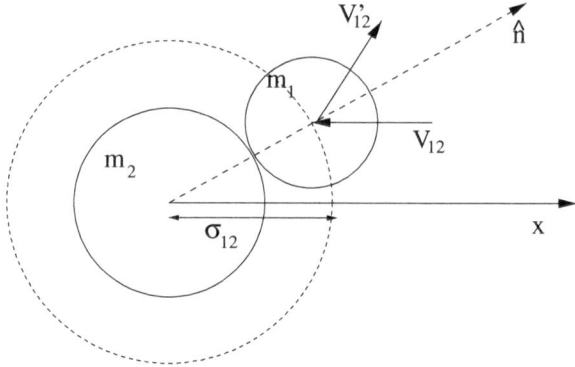

Fig. 2.2 The collision between two elastic smooth hard spheres

$$\chi = 2 \arccos \left(\frac{b}{\sigma_{12}} \right) \tag{2.13}$$

and the dependence from the initial relative velocity disappears: only geometry determines the deflection angle.

In the study of *smooth* hard spheres (i.e. such that particles' rotation is not relevant), a complete description of the dynamics requires only the positions of the centers **r** and their velocities **v**. In particular the collision is an instantaneous transformation of the velocities of two particles i and j at contact which are "reflected" with the following rule (see Fig. 2.2):

$$\mathbf{v}'_i = \mathbf{v}_i - \frac{2m_2}{m_1 + m_2} \hat{\mathbf{n}} [\hat{\mathbf{n}} \cdot (\mathbf{v}_i - \mathbf{v}_j)] \tag{2.14}$$

$$\mathbf{v}'_j = \mathbf{v}_j + \frac{2m_1}{m_1 + m_2} \hat{\mathbf{n}} [\hat{\mathbf{n}} \cdot (\mathbf{v}_i - \mathbf{v}_j)] \tag{2.15}$$

where $\hat{\mathbf{n}} = (\mathbf{r}_i - \mathbf{r}_j)/|\mathbf{r}_i - \mathbf{r}_j|$ and the primes denote the velocities after the collision. This collision rule conserves momentum and kinetic energy. It only changes the direction of the component of the relative velocity of the particles in the direction of $\hat{\mathbf{n}}$ (normal component), leaving unchanged the tangential component.

2.1.2 Statistics of Hard Spheres Collisions

The concept of *mean free path* was introduced in 1858 by Rudolf Clausius [17] and paved the road to the kinetic theory of gases. For the sake of simplicity, here I consider a single species gas composed of elastic smooth hard spheres, all having the same diameter σ and mass m (see [16]).

The *mean free time* is the average time between two successive collisions of a single particle. I define $\omega_c dt$ the probability that a given particle suffers a collision between time t and $t + dt$ (ω_c is called collision frequency) and assume that ω_c is independent of the past collisional history of the particle. The probability $f_{time}dt$ of having a free time between two successive collisions larger than t and shorter than $t + dt$ is equal to the product of the probability that no collision occurs in the time interval $[0, t]$ and the probability that a collision occurs in the interval $[t, t + dt]$:

$$f_{time}(t)dt = P_{time}(t)\omega_c dt, \qquad (2.16)$$

where $P_{time}(t)$ is the survival probability, that is the probability that no collisions happen between 0 and t, and can be calculated observing that $P_{time}(t + dt) = P_{time}(t)P_{time}(dt) = P_{time}(t)(1 - \omega_c dt)$ so that $dP_{time}/dt = -\omega_c P_{time}$, i.e. $P_{time}(t) = e^{-\omega_c t}$.

Finally one can calculate the average of the free time using the probability density $f_{time}(t)$:

$$\tau_c = \int_0^\infty dt\, t f_{time}(t) = \int_0^\infty dt\, t \omega_c e^{-\omega_c t} = \frac{1}{\omega_c}. \qquad (2.17)$$

With the same sort of calculations an expression for the mean free path, that is the average distance traveled by a particle between two successive collisions, can be calculated. One again assumes that there is a well defined quantity (independent of the collisional history of the particle) αdl which is the probability of a collision during the travel between distances l and $l + dl$. The survival probability in terms of space traveled is $P_{path}(l) = e^{-\alpha l}$ and the probability density of having a free distance l is $f_{path}(l) = e^{-\alpha l}\alpha$ so that the mean free path is given by:

$$\lambda = \frac{1}{\alpha} \qquad (2.18)$$

Above, for simplicity, I have considered a homogeneous probability for collisions. A more precise treatment requires to consider the hard core collision process as a non-homogeneous stochastic Poisonnian process: indeed the transition rates for the particle's change of velocity depend on the relative velocity between the colliding particles [50]. This is discussed in details in Chap. 4.

The other important statistical quantity in the study of binary collisions is the so-called *differential scattering cross section s*. In a unit time a particle suffers a number of collisions which can be seen as the incidence of fluxes of particles coming with different approaching velocities $\mathbf{V}_{12}$ and scattered to new different departure velocities $\mathbf{V}'_{12}$. Given a certain approaching velocity $\mathbf{V}_{12}$ the incident particles arrive with slightly different impact parameters (due to the extension of the particles) and therefore are scattered in a solid angle $d\Omega'$. If I_0 denotes the intensity of the beam of particles that come with an average approaching speed $\mathbf{V}_{12}$, which is the number of particles intersecting in unit time a unit area perpendicular to the beam ($I_0 = n V_{12}$ with n the number density of the particles), then the rate of scattering dR into the small solid angle element $d\Omega'$ is given by

$$\frac{dR}{d\Omega'} = I_0 s(\mathbf{V}_{12}, \mathbf{V}'_{12}) \tag{2.19}$$

where s is a factor of proportionality with the dimensions of an area (in $3D$) which is called differential cross section and depends on the relative velocity vectors before and after the collisions. The total rate of particles scattered in all directions, R is the integral of the last equation:

$$R = I_0 \int\!\!\!\int_{4\pi} d\Omega' s(\mathbf{V}_{12}, \mathbf{V}'_{12}) = S I_0 \tag{2.20}$$

and defines the total scattering cross section S.

In the case of a spherically symmetric central field of force, the differential cross section is a function only of the modulus of the initial relative velocity V_{12}, the angle of deflection χ, and the impact parameter b which in turn, once fixed the potential $U(r)$, is a function only of χ and V_{12}, that is $s = s(V_{12}, \chi)$. In particular it is easily seen that

$$s(V_{12}, \chi) = -\frac{b(V_{12}, \chi)}{\sin \chi} \frac{db}{d\chi}. \tag{2.21}$$

The differential scattering cross section for hard spheres is calculated from Eq. (2.21) obtaining a very simple formula: $s(V_{12}, \chi) = \sigma^2/4$ which can be integrated over the entire solid angle space giving an expression for the total cross section $S = \pi\sigma^2$. This result is consistent with the physical intuition of the cross section: it is the average of the areas of influence of the scatterer in the planes perpendicular to the approaching velocities of the incident particles.

To conclude this paragraph I recall that the collision frequency in a homogeneous stationary gas is related to the total scattering cross section by the relation

$$\omega_c = nS\langle V_{12} \rangle \tag{2.22}$$

where n is the uniform density of the gas and $\langle V_{12} \rangle$ is an average of the relative velocities. Assuming that velocities in the gas are independent and their distribution is the Maxwell-Boltzmann distribution:

$$P(\mathbf{v}) = \frac{m^{3/2}}{(2\pi k_B T)^{3/2}} e^{-\frac{mv^2}{2k_B T}} \tag{2.23}$$

the collision frequency can be calculated obtaining the formula:

$$\omega_c = \frac{2\sqrt{2}}{\sqrt{\pi}} n S v_T \tag{2.24}$$

where v_T is defined as

$$v_T = \sqrt{\frac{2k_B T}{m}}. \tag{2.25}$$

In the same way the mean free path is given by

$$\lambda = \frac{1}{\sqrt{2}nS}. \tag{2.26}$$

2.1.3 Inelasticity

Granular particles collide dissipating the kinetic energy of their relative motion [11]. This is due to the macroscopic nature of the grains: during the interaction, irreversible processes happen inside the grain and energy is dissipated in form of heat. In a collision between two free particles, these processes conserve momentum so that the velocity of the center of mass of the two grains is not modified.

Many models of the binary inelastic collision have been proposed (soft spheres [13, 25, 35, 51, 52] as well as hard spheres models [12, 21, 27, 41]): this is usually a relatively difficult problem. Simplification often pays more, as very idealized models lead to physically meaningful results. The most used model in granular gas literature is also the simplest: the gas of inelastic smooth hard spheres, with *fixed restitution coefficient*. It is given by the following prescriptions:

$$m_1 \mathbf{v}'_1 + m_2 \mathbf{v}'_2 = m_1 \mathbf{v}_1 + m_2 \mathbf{v}_2 \tag{2.27a}$$
$$(\mathbf{v}'_1 - \mathbf{v}'_2) \cdot \hat{\mathbf{n}} = -r(\mathbf{v}_1 - \mathbf{v}_2) \cdot \hat{\mathbf{n}} \tag{2.27b}$$

where, as usual, the primes denote the postcollisional velocities, $\hat{\mathbf{n}}$ is the unit vector in the direction joining the centers of the grains, and $0 \leq r \leq 1$. In this model the collisions happen at contact and are instantaneous. When $r = 1$ the gas is elastic and the rule coincides with the collision description for hard spheres given in the Sect. 2.1.1. When $r = 0$ the gas is perfectly inelastic, that is the particles exit from the collision with no relative velocity in the $\hat{\mathbf{n}}$ direction.

As a matter of fact, the transformation that gives the (primed) postcollisional velocities from the precollisional velocities of the two colliding particles is

$$\mathbf{v}'_1 = \mathbf{v}_1 - (1+r)\frac{m_2}{m_1 + m_2}((\mathbf{v}_1 - \mathbf{v}_2) \cdot \hat{\mathbf{n}})\hat{\mathbf{n}} \tag{2.28a}$$
$$\mathbf{v}'_2 = \mathbf{v}_2 + (1+r)\frac{m_1}{m_1 + m_2}((\mathbf{v}_1 - \mathbf{v}_2) \cdot \hat{\mathbf{n}})\hat{\mathbf{n}} \tag{2.28b}$$

Sometimes it may be useful to have the reverse transformation that give precollisional velocities from postcollisional ones, with the primes exchanged:

$$\mathbf{v}_1' = \mathbf{v}_1 - \left(1 + \frac{1}{r}\right) \frac{m_2}{m_1 + m_2} ((\mathbf{v}_1 - \mathbf{v}_2) \cdot \hat{\mathbf{n}})\hat{\mathbf{n}} \tag{2.29a}$$

$$\mathbf{v}_2' = \mathbf{v}_2 + \left(1 + \frac{1}{r}\right) \frac{m_1}{m_1 + m_2} ((\mathbf{v}_1 - \mathbf{v}_2) \cdot \hat{\mathbf{n}})\hat{\mathbf{n}} \tag{2.29b}$$

As it can be seen, the inverse transformation is equivalent to a change of the restitution coefficient $r \to 1/r$. Obviously, in the case of a perfectly inelastic gas ($r = 0$) there is no inverse transformation. I also note that in 1D and when $m_1 = m_2$ Eqs. (2.28a, 2.28b) become:

$$v_1' = \frac{1 - r}{2} v_1 + \frac{1 + r}{2} v_2 \tag{2.30a}$$

$$v_2' = \frac{1 + r}{2} v_1 + \frac{1 - r}{2} v_2 \tag{2.30b}$$

which correspond to an exact exchange of velocities in the elastic ($r = 1$) case, and in a sticky collision in the perfectly inelastic ($r = 0$) case. In dimensions higher than *one* the $r = 0$ case is very different from the so-called *sticky gas*, which is defined as a gas of hard spheres that in a collision become stuck together. In one dimension, instead, the $r = 0$ case may be considered equivalent to a sticky gas but a further prescription of "stickiness" must be given in order to consider collisions among more than two particles.

Variants of this models have been largely used in the literature. The importance of tangential frictional forces acting on the grains at contact may be studied taking into account the rotational degree of freedom of the particles, i.e. adding a variable ω_i to each grain. The simplest model which takes into account the rotational degree of freedom of particles is the rough hard spheres gas [22, 28, 31, 36–38, 42]. In this model the postcollisional translational and angular velocities are given by the following equations (where the bottom signs in $\pm$ or $\mp$ are to be considered for particle 2):

$$\mathbf{v}_{1,2}' = \mathbf{v}_{1,2} \mp \frac{1 + r}{2} \mathbf{v}_n \mp \frac{q(1 + \beta)}{2q + 2} (\mathbf{v}_t + \mathbf{v}_r) \tag{2.31a}$$

$$\boldsymbol{\omega}_{1,2}' = \boldsymbol{\omega}_{1,2} + \frac{1 + \beta}{\sigma(1 + q)} [\hat{\mathbf{n}} \times (\mathbf{v}_t + \mathbf{v}_r)] \tag{2.31b}$$

where q is the dimensionless moment of inertia defined by $I = qm(\sigma/2)^2$ (with I the moment of inertia of the hard object), e.g. $q = 1/2$ for disks and $q = 2/5$ for spheres; $\mathbf{v}_n = ((\mathbf{v}_1 - \mathbf{v}_2) \cdot \hat{\mathbf{n}})\hat{\mathbf{n}}$ is the normal relative velocity component, $\mathbf{v}_t = \mathbf{v}_1 - \mathbf{v}_2 - \mathbf{v}_n$ is the tangential velocity component due to translational motion, while $\mathbf{v}_r = -\sigma(\boldsymbol{\omega}_1 - \boldsymbol{\omega}_2)$ is the tangential velocity component due to particle rotation. In Eqs. (2.31a, 2.31b) the tangential restitution coefficient β appears: it may take any value between -1 and $+1$. When $\beta = -1$ tangential effects disappear, i.e. rotation is not affected by collision (rough spheres become smooth spheres). When

$\beta = +1$ the particles are said to have perfectly rough surface. It can be easily seen that (when $r = 1$) energy is conserved for $\beta = \pm 1$.

Other models for collisions have been introduced, justified by a deeper analysis of the collision process. In these models the restitution coefficient r (or the coefficients r and β in the more detailed description given above) depends on the relative velocity of the colliding particles. In particular it has been seen that the collision tends to become more and more elastic as the relative velocity tends to zero. This refined description, referred to as 'viscoelastic' model [10, 26], has relevance in different issues of the statistical mechanics of granular gases. An important kinetic instability of the cooling (and sometimes driven) granular gases is the so-called *inelastic collapse* [40, 41], i.e. a divergence of the local collision rate due to the presence of a few particles trapped very close to each other: simulations of the gas with the viscoelastic model have shown that this instability is removed, suggesting that it is an artifact of the fixed restitution coefficient idealization.

Here, I give an expression of the leading term for the velocity dependence of the normal restitution coefficient r in the viscoelastic model (the viscoelastic theory may be applied to give also a velocity dependent expressions for the tangential restitution coefficient):

$$r = 1 - C_1 |(\mathbf{v}_1 - \mathbf{v}_2) \cdot \hat{\mathbf{n}}|^{1/5} + \cdots \tag{2.32a}$$

where C_1 depends on the physical properties of the spheres (mass, density, radius, Young modulus, viscosity).

2.1.4 Inelastic Collapse

In the 1990s, several numerical studies have unveiled a problem in the model of inelastic collisions with a fixed restitution coefficient. Such a problem went under the name of "inelastic collapse". The simplest example involves just three particles on a line, as shown in Fig. 2.3 [5, 40]. The two outer particles move monotonically toward each other and the one in the middle bounces between them. One can easily show that, after the two collisions shown in the figure, the relation between the final and initial velocities is $\mathbf{u}' = \mathcal{M} \mathbf{u}$ where $\mathbf{u} = (v_1, v_2, v_3)^T$ and $\mathcal{M}$ is a 3x3 matrix whose entries are quadratic polynomials in r. If this matrix has one real eigenvalue in the interval $(0, 1)$, the cycle shown in Figure endlessly repeats with geometrically smaller space and time scales at each successive cycle. This requires $r = r_c < 7 - 4\sqrt{3} \approx 0.0718$ to happen. In this case an infinite number of collision happens in a finite time. When $r > r_c$, inelastic collapse can still occur but with the collective participation of more than three particles or with the presence of an inelastic wall (because of symmetry, this is equivalent to an interaction between four inelastic particles), as displayed in the Figure. As the coefficient of restitution r increases toward 1, the number of particles required for collapse increases. For instance, with $r = 0.8$, it is required that $N = 16$

Fig. 2.3 Examples of particles' trajectories with or without a wall: **a** three particles collapse $(r < 7 - 4\sqrt{3} \approx 0.0718)$; **b** two particles bouncing off an inelastic wall: when $r > 0.346015$ they finally leaves the wall and never come back; **c** critical value $r = 0.346015$, the inner ball remains stationary after two collisions with the other particle; **d** when $r < 3 - 2\sqrt{2} \approx 0.17157$ there is inelastic collapse. Reproduced with permission from McNamara and Young [40]. Copyright 1992, AIP Publishing LLC

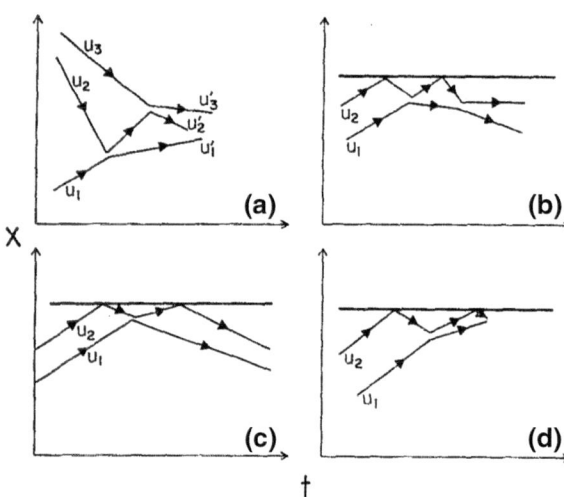

particles bounce off an inelastic wall. Rough estimates suggest (in agreement with numerical calculations) that $N_{\min}(r) \approx ln(4/(1 - r))/(1 - r)$ as $r \to 1$.

In more than 1 dimension, the trapping necessary to have collapse can be realized in a large cluster, as shown in Fig. 2.4.

Fig. 2.4 A snapshot from a MD simulation of cooling inelastic *hard spheres*. The particles in *black* are those that have participated in the last collisions, just before a collapse. Reprinted with permission from Schorghofer and Zhou [48]. Copyright 1996 by the American Physical Society

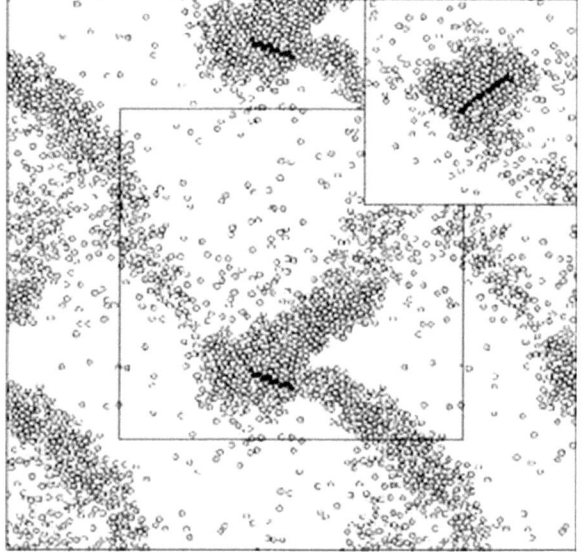

2.2 The Boltzmann Equation

The Boltzmann equation for a gas of elastic or inelastic hard spheres can be derived in several ways [15]. Here, I review the typical reduction scheme which starts from the Liouville equation and goes through the BBGKY hierarchy.

2.2.1 Liouville and Pseudo-Liouville Equations

In order to discuss the behavior of a system of N identical hard spheres (of diameter σ and mass m) it is natural to introduce the phase space, i.e., a $6N$—dimensional space where the coordinates are the $3N$ components of the N position vectors of the sphere centers $\mathbf{r}_i$ and the $3N$ components of the N velocities $\mathbf{v}_i$. The state of the system is represented by a point in this space. I call $\mathbf{z}$ the $6N$-dimensional position vector of this point. If the positions $\mathbf{r}_i$ of the spheres are restricted in a space region Ω, then the full phase space $\mathbf{D}$ is given by the product $\Omega^N \times \Re^{3N}$

If the state is not known with absolute accuracy, one must introduce a probability density $P(z, t)$ which is defined by

$$Prob(z \in \mathbf{D} \quad \text{at time } t) = \int_{\mathbf{D}} P(\mathbf{z}, t)d\mathbf{z} \tag{2.33}$$

where $d\mathbf{z}$ is the Lebesgue measure in phase space. One implicitly assumes that the probability is a measure absolutely continuous with respect to the Lebesgue measure. The mean value of a dynamical observable $A(\mathbf{z})$ can be calculated from either the following expressions:

$$\int_{\infty} d\mathbf{z}P(\mathbf{z}, 0)A(\mathbf{z}(t)) = \int_{\infty} d\mathbf{z}P(\mathbf{z}, t)A(\mathbf{z}) \tag{2.34}$$

which are respectively the Lagrangian and Eulerian averages (analogous to the Heisenberg and Schroedinger averages in quantum mechanics). In Eq. (2.34) the time dependence of the observable A and of the distribution P is due to the time evolution operator S_t, also called *streaming operator*, that is $A(\mathbf{z}(t)) \equiv S_t(\mathbf{z})A(\mathbf{z})$. Considering the equivalence in Eq. (2.34) as an inner product implies that

$$P(\mathbf{z}, t) = S_t^{\dagger} P(\mathbf{z}, 0) \tag{2.35}$$

where $S_t^{\dagger}$ is the adjoint of S_t.

In a general system (not necessarily made of hard spheres) with conservative and additive interactions, the force between the particle pair (ij) is $\mathbf{F}_{ij} = -\partial U(r_{ij})/\partial \mathbf{r}_{ij}$ so that the time evolution operator is given by:

$$S_t(\mathbf{z}) = \exp[t L(\mathbf{z})] = \exp\left[t \sum_i L_i^0 - t \sum_{i<j} \Theta(ij) \right] \qquad (2.36)$$

where the *Liouville operator* $L(\mathbf{z})\ldots \equiv \{H(\mathbf{z}), \ldots\}$ is the Poisson bracket with the Hamiltonian, so that

$$L_i^0 = \mathbf{v}_i \cdot \frac{\partial}{\partial \mathbf{r}_i} \qquad (2.37a)$$

$$\Theta(ij) = \frac{1}{m} \frac{\partial U(r_{ij})}{\partial \mathbf{r}_{ij}} \cdot \left(\frac{\partial}{\partial \mathbf{v}_i} - \frac{\partial}{\partial \mathbf{v}_j} \right) \qquad (2.37b)$$

and $S_t(\mathbf{z})$ is a unitary operator, $S_t^\dagger = S_{-t}$, while $L^\dagger = -L$. In Eq. (2.36) the evolution operator S_t has been divided into a free streaming operator $S_t^0 = \exp[t \sum_i L_i^0]$ which generates the free particle trajectories, plus a term containing the binary interactions among the particles.

Finally the Liouville equation is obtained writing explicitly Eq. (2.35):

$$\frac{\partial}{\partial t} P(\mathbf{z}, t) = \left(-\sum_i L_i^0 + \sum_{i<j} \Theta(ij) \right) P(\mathbf{z}, t) \qquad (2.38)$$

which is an expression of the incompressibility of the flow in phase space.

In the specific case of identical hard spheres, the interaction among particles is defined by Eqs. (2.12a, 2.12b). It can be shown that this kind of interaction carries no contraction of phase space at collision, i.e.

$$P(\mathbf{z}', t) = P(\mathbf{z}, t) \qquad (2.39)$$

where $\mathbf{z}'$ and $\mathbf{z}$ are the phase space points before and after a collision. This can be considered a form of detailed balance law. It is important to stress that $\mathbf{z}' \neq \mathbf{z}$: a collision represents a time discontinuity in the velocity section of phase space. In particular I use the elastic collision model defined in this list of prescriptions [it coincides with the collision rule for smooth hard spheres, see Eq. (2.14)]:

$$|\mathbf{r}_i - \mathbf{r}_j| = \sigma \qquad (2.40a)$$

$$\hat{\mathbf{n}}_{ij} = (\mathbf{r}_i - \mathbf{r}_j)/\sigma \qquad (2.40b)$$

$$\mathbf{V}_{ij} = \mathbf{v}_i - \mathbf{v}_j \qquad (2.40c)$$

$$\mathbf{V}_{ij} \cdot \hat{\mathbf{n}}_{ij} < 0 \qquad (2.40d)$$

$$\mathbf{z} \equiv (\mathbf{r}_1, \mathbf{v}_1, \mathbf{r}_2, \mathbf{v}_2, \ldots, \mathbf{r}_i, \mathbf{v}_i, \ldots, \mathbf{r}_j, \mathbf{v}_j, \ldots, \mathbf{r}_N, \mathbf{v}_N) \qquad (2.40e)$$

$$\mathbf{z}' \equiv (\mathbf{r}_1, \mathbf{v}_1, \mathbf{r}_2, \mathbf{v}_2, \ldots, \mathbf{r}_i', \mathbf{v}_i', \ldots, \mathbf{r}_j', \mathbf{v}_j', \ldots, \mathbf{r}_N, \mathbf{v}_N) \qquad (2.40f)$$

$$\mathbf{r}_i' = \mathbf{r}_i \qquad (2.40g)$$

$$\mathbf{r}_j' = \mathbf{r}_j \qquad (2.40h)$$

$$\mathbf{v}_i' = \mathbf{v}_i - \hat{\mathbf{n}}_{ij}(\hat{\mathbf{n}}_{ij} \cdot \mathbf{V}_{ij}) \tag{2.40i}$$

$$\mathbf{v}_j' = \mathbf{v}_j + \hat{\mathbf{n}}_{ij}(\hat{\mathbf{n}}_{ij} \cdot \mathbf{V}_{ij}) \tag{2.40j}$$

these relations conserve the total momentum and the total energy of the system.

In order to derive the Boltzmann equation, the collisions events $\mathbf{z} \rightarrow \mathbf{z}'$ are considered as boundary conditions and the Liouville Equation (2.38) is restricted to the interior of the phase space region $\Lambda \equiv \Omega^N \times \Re^{3N} - \Lambda_{ov}$ where

$$\Lambda_{ov} = \left\{ \mathbf{z} \in \Omega^N \times \Re^{3N} \mid \exists i, j \in \{1, 2, \ldots, N\} \, (i \neq j) : |\mathbf{r}_i - \mathbf{r}_j| < \sigma \right\} \tag{2.41}$$

is the set of phase space points such that one or more pairs of spheres are overlapping. With this conditions, the Liouville equation reads:

$$\frac{\partial}{\partial t} P(\mathbf{z}, t) = \left(-\sum_i \mathbf{v}_i \cdot \frac{\partial}{\partial \mathbf{r}_i} \right) P(\mathbf{z}, t) \quad (\mathbf{z} \in \Lambda) \tag{2.42a}$$

$$P(\mathbf{z}, t) = P(\mathbf{z}', t) \quad (\mathbf{z} \in \partial \Lambda). \tag{2.42b}$$

This version of the Liouville equation is time-discontinuous: this means that formal perturbation expansions used in usual many-body theory methods cannot be applied.

An alternative master equation for the probability density function in the phase space can be derived [18]. The streaming operator S_t for hard spheres is not defined for any point of the phase space $\mathbf{z} \in \Lambda_{ov}$. In the calculation of the average (2.34) of physical observables, this is not a problem, as the streaming operators appears multiplied by $P(\mathbf{z}, 0)$ which is proportional to the characteristic function $X(\mathbf{z})$ of the set Λ (the characteristic function is 1 for points belonging to the set and 0 for points outside of it). In perturbation expansions it is safer to have a streaming operator defined for every point of the configurational space. A standard representation, defined for all points in the phase space, has been developed for elastic hard spheres and is based on the binary collision expansion of $S_t(\mathbf{z})$ in terms of binary collision operators. The binary collision operator is defined in terms of two-body dynamics through the following representation of the streaming operator for the evolution of two particles:

$$S_t(1, 2) = S_t^0(1, 2) + \int_0^t d\tau \, S_\tau^0(1, 2) T_+(1, 2) S_{t-\tau}^0(1, 2), \tag{2.43}$$

with $S_t^0 = \exp(t L_0)$ the free flow operator and a collision operator

$$T_+(1, 2) = \sigma^2 \int_{\mathbf{V}_{12} \cdot \hat{\mathbf{n}} < 0} d\hat{\mathbf{n}} |\mathbf{V}_{12} \cdot \hat{\mathbf{n}}| \delta(\sigma \hat{\mathbf{n}} - (\mathbf{r}_1 - \mathbf{r}_2))(b_c - 1), \tag{2.44}$$

where b_c is a substitution operator that replaces $\mathbf{v}_1$, $\mathbf{v}_2$ with $\mathbf{v}_1'$, $\mathbf{v}_2'$ (see Eqs. (2.40a)).

The Eq. (2.43) is a representation of the evolution of two particles as a convolution of free flow and collisional events. Noting that $T_+(1, 2)S_\tau^0(1, 2)T_+(1, 2) = 0$ for $\tau > 0$ (two hard spheres cannot collide more than once), Eq. (2.43) can be put in the form

$$S_t(1, 2) = \exp\left\{t[L_0(1, 2) + T_+(1, 2)]\right\}, \qquad (2.45)$$

that can be generalized to the N-particle streaming operator (here considered for the case of an infinite volume):

$$S_{\pm t}(\mathbf{z}) = \exp\left\{\pm t\left[L_0(\mathbf{z}) \pm \sum_{i<j} T_{\pm}(i, j)\right]\right\} \qquad (2.46)$$

where

$$T_-(1, 2) = \sigma^2 \int_{\mathbf{V}_{12}\cdot\hat{\mathbf{n}}>0} d\hat{\mathbf{n}}|\mathbf{V}_{12}\cdot\hat{\mathbf{n}}|\delta(\mathbf{r}_1 - \mathbf{r}_2 - \sigma\hat{\mathbf{n}})(b_c - 1). \qquad (2.47)$$

Equation (2.46) defines the so-called *pseudo-streaming operator*. In order to write an analogue of the Liouville Equation (2.38), the adjoint of $S_{\pm t}$ is needed; its definition is identical to that in Eq. (2.46) but for the binary collision operators which must be replaced by their adjoints:

$$\overline{T}_{\pm}(1, 2) = \sigma^2 \int_{\mathbf{V}_{12}\cdot\hat{\mathbf{n}}\underset{>}{<}0} d\hat{\mathbf{n}}|\mathbf{V}_{12}\cdot\hat{\mathbf{n}}|[\delta(\mathbf{r}_1 - \mathbf{r}_2 - \sigma\hat{\mathbf{n}})b_c - \delta(\mathbf{r}_1 - \mathbf{r}_2 + \sigma\hat{\mathbf{n}})]. \quad (2.48)$$

Finally the pseudo-Liouville equation can be written:

$$\frac{\partial}{\partial t}P(\mathbf{z}, t) = \left(-\sum_i L_i^0 + \sum_{i<j} \overline{T}_-(ij)\right)P(\mathbf{z}, t). \qquad (2.49)$$

This equation is the analogue of Eq. (2.38) for the case of hard core potential (hard spheres). In this sense it replaces Eqs. (2.42a, 2.42b) and its modification for inelastic collisions will be discussed in Sect. 2.3.

2.2.2 The BBGKY Hierarchy

Reduced (marginal) probability densities P_s are defined as

$$P_s(\mathbf{r}_1, \mathbf{v}_1, \mathbf{r}_2, \mathbf{v}_2, \ldots, \mathbf{r}_s, \mathbf{v}_s, t) = \int\limits_{\Omega^{N-s} \times \mathfrak{R}^{3(N-s)}} P(\mathbf{r}_1, \mathbf{v}_1, \mathbf{r}_2, \mathbf{v}_2, \ldots, \mathbf{r}_N, \mathbf{v}_N, t) \prod_{j=s+1}^{N} d\mathbf{r}_j d\mathbf{v}_j.$$

(2.50)

In order to derive an evolution equation for P_s the first step is to integrate Eqs. (2.42a, 2.42b) with respect to the variables $\mathbf{r}_j$ and $\mathbf{v}_j$ $(s+1 \leq j \leq N)$ over $\Omega^{N-s} \times \mathfrak{R}^{3(N-s)}$, obtaining:

$$\frac{\partial P_s}{\partial t} + \sum_{i=1}^{s} \int\limits_{\Lambda_s} \mathbf{v}_i \cdot \frac{\partial P}{\partial \mathbf{r}_i} \prod_{j=s+1}^{N} d\mathbf{r}_j d\mathbf{v}_j + \sum_{k=s+1}^{N} \int\limits_{\Lambda_s} \mathbf{v}_k \cdot \frac{\partial P}{\partial \mathbf{r}_k} \prod_{j=s+1}^{N} d\mathbf{r}_j d\mathbf{v}_j = 0, \quad (2.51)$$

where the integration space Λ_s extends to the entire $\mathfrak{R}^{3(N-s)}$ for the velocity variables, while it extends to Ω^{N-s} deprived of the spheres $|\mathbf{r}_i - \mathbf{r}_j| < \sigma$ $(i = 1, \ldots, N, i \neq j)$ with respect to the position variables.

The typical term in the first sum contains the integral of a derivative with respect to a variable $\mathbf{r}_i$ over which one does not integrate, but in the exchange of order between integration and derivation one must take into account the domain boundaries which depend on $\mathbf{r}_i$, writing:

$$\int\limits_{\Lambda_s} \mathbf{v}_i \cdot \frac{\partial P}{\partial \mathbf{r}_i} \prod_{j=s+1}^{N} d\mathbf{r}_j d\mathbf{v}_j = \mathbf{v}_i \cdot \frac{\partial P_s}{\partial \mathbf{r}_i} - \sum_{k=s+1}^{N} \int\limits_{\Lambda_s} P_{s+1} \mathbf{v}_i \cdot \hat{\mathbf{n}}_{ik} d\sigma_{ik} d\mathbf{v}_k \quad (2.52)$$

where $\hat{\mathbf{n}}_{ik}$ is the outer normal to the sphere $|\mathbf{r}_i - \mathbf{r}_k| = \sigma$, $d\sigma_{ik}$ is the surface element on the same sphere and P_{s+1} has k as its $(s+1) - th$ index.

The typical term in the second sum in Eq. (2.51) can be immediately integrated by means of the Gauss theorem, since it involves the integration of a derivative taken with respect to one of the integration variables (and assuming that the boundary of Ω is a specular reflecting wall or a periodical boundary condition):

$$\int\limits_{\Lambda_s} \mathbf{v}_k \cdot \frac{\partial P}{\partial \mathbf{r}_k} \prod_{j=s+1}^{N} d\mathbf{r}_j d\mathbf{v}_j = \sum_{i=1}^{s} \int P_{s+1} \mathbf{v}_k \cdot \hat{\mathbf{n}}_{ik} d\sigma_{ik} d\mathbf{v}_k$$

$$+ \sum_{i=s+1, i \neq k}^{N} \int P_{s+2} \mathbf{v}_k \cdot \hat{\mathbf{n}}_{ik} d\sigma_{ik} d\mathbf{v}_k d\mathbf{r}_i d\mathbf{v}_i. \quad (2.53)$$

The last term in the above equation, when summed over $s + 1 \leq k \leq N$ vanishes: this fact directly stems from the equivalence Eq. (2.42b). Moreover, in both above equations the integral containing the term P_{s+1} is the same no matter what the value of the dummy index k is, so that I can drop the index and write $\mathbf{r}_*, \mathbf{v}_*$ instead of $\mathbf{r}_k, \mathbf{v}_k$.

As a matter of fact, Eq. (2.51) finally reads:

$$\frac{\partial P_s}{\partial t} + \sum_{i=1}^{s} \mathbf{v}_i \cdot \frac{\partial P_s}{\partial \mathbf{r}_i} = (N-s) \sum_{i=1}^{s} \int P_{s+1} \mathbf{V}_i \cdot \hat{\mathbf{n}}_i d\sigma_i d\mathbf{v}_* \qquad (2.54)$$

where $\mathbf{V}_i = \mathbf{v}_i - \mathbf{v}_*$, $\hat{\mathbf{n}}_i = (\mathbf{r}_i - \mathbf{r}_*)/\sigma$ and the arguments of P_{s+1} are $(\mathbf{r}_1, \mathbf{v}_1, \mathbf{r}_2, \mathbf{v}_2,$ $\ldots, \mathbf{r}_s, \mathbf{v}_s, \mathbf{r}_*, \mathbf{v}_*, t)$. Integrations in Eq. (2.54) are performed over the 1-particle velocity space $\mathfrak{R}^3$ and over the sphere S^i (given by the condition $|\mathbf{r}_i - \mathbf{r}_*| = \sigma$) with surface elements $d\sigma_i$. Eq. (2.54) is complemented by reflecting boundary conditions (of the same kind of (2.40a)) on the reduced boundary surface Λ_s.

Equation (2.54) states that the evolution of the reduced probability density P_s is governed by the free evolution operator of the s-particles dynamics, which appears in the left hand side, with corrections due to the effect of the interaction with the remaining $(N-s)$ particle. The effect of this interaction is described by the right-hand side of this equation.

Usually Eq. (2.54) is written in a different form, obtained using some symmetries of the problem. In particular one can separate the sphere S^i of integration in the right-hand side, in the two hemispheres S^i_+ and S^i_- defined respectively by $\mathbf{V_i} \cdot \hat{\mathbf{n}}_i > 0$ and $\mathbf{V_i} \cdot \hat{\mathbf{n}}_i < 0$ (considering also that $d\sigma_i = \sigma^2 d\hat{\mathbf{n}}_i$):

$$\int P_{s+1} \mathbf{V}_i \cdot \hat{\mathbf{n}}_i d\sigma_i d\mathbf{v}_* = \sigma^2 \int_{\mathfrak{R}^3} \int_{S^i_+} P_{s+1} |\mathbf{V}_i \cdot \hat{\mathbf{n}}_i| d\hat{\mathbf{n}}_i d\mathbf{v}_* - \sigma^2 \int_{\mathfrak{R}^3} \int_{S^i_-} P_{s+1} |\mathbf{V}_i \cdot \hat{\mathbf{n}}_i| d\hat{\mathbf{n}}_i d\mathbf{v}_*,$$
$$(2.55)$$

and observe that in the S^i_+ integration are included all phase space points such that particle i and particle $*$ (the $(s+1)-th$ generic particle) are coming out from a collision: this means that on the sphere S^i_+ I can write the substitution

$$P_{s+1}(\mathbf{r}_1, \mathbf{v}_1, \ldots, \mathbf{r}_i, \mathbf{v}_i, \ldots \mathbf{r}_s, \mathbf{v}_s, \mathbf{r}_i - \sigma \hat{\mathbf{n}}_i, \mathbf{v}_*)$$
$$\rightarrow P_{s+1}(\mathbf{r}_1, \mathbf{v}_1, \ldots, \mathbf{r}_i, \mathbf{v}_i - \hat{\mathbf{n}}_i(\hat{\mathbf{n}}_i \cdot \mathbf{V}_i), \ldots \mathbf{r}_s, \mathbf{v}_s, \mathbf{r}_i - \sigma \hat{\mathbf{n}}_i, \mathbf{v}_* + \hat{\mathbf{n}}_i(\hat{\mathbf{n}}_i \cdot \mathbf{V}_i)).$$
$$(2.56)$$

Moreover one can make the change of variable in the second integral (that on the sphere S^i_-) $\hat{\mathbf{n}}_i \rightarrow -\hat{\mathbf{n}}_i$ which only changes the integration range $S^i_- \rightarrow S^i_+$. Finally, replacing $\hat{\mathbf{n}}_i$ with simply $\hat{\mathbf{n}}$ (and therefore $S^i_+ \rightarrow S_+$) one has:

$$\frac{\partial P_s}{\partial t} + \sum_{i=1}^{s} \mathbf{v}_i \cdot \frac{\partial P_s}{\partial \mathbf{r}_i} = (N-s)\sigma^2 \sum_{i=1}^{s} \int_{\mathfrak{R}^3} \int_{S_+} (P'_{s+1} - P_{s+1})|\mathbf{V}_i \cdot \hat{\mathbf{n}}| d\hat{\mathbf{n}} d\mathbf{v}_*, \quad (2.57)$$

where I have defined

$$P'_{s+1} = P_{s+1}(\mathbf{r}_1, \mathbf{v}_1, \ldots, \mathbf{r}_i, \mathbf{v}_i - \hat{\mathbf{n}}_i(\hat{\mathbf{n}}_i \cdot \mathbf{V}_i), \ldots \mathbf{r}_s, \mathbf{v}_s, \mathbf{r}_i - \sigma \hat{\mathbf{n}}_i, \mathbf{v}_* + \hat{\mathbf{n}}_i(\hat{\mathbf{n}}_i \cdot \mathbf{V}_i)).$$
$$(2.58)$$

The system of Eqs. (2.57) is usually called the BBGKY hierarchy for the hard sphere gas (from Bogoliubov, Born, Green, Kirkwood and Yvon, sometimes called simply Bogoliubov hierarchy).

2.2.3 The Boltzmann Hierarchy and the Boltzmann Equation

In a rarefied gas, N is a very large number and σ is very small; let us say, to fix ideas, that we have a box whose volume is $1\,\text{cm}^3$ at room temperature and atmospheric pressure. Then $N \simeq 10^{20}$ and $\sigma \simeq 10^{-8}\text{cm}$ and [from Eq. (2.57)] for small s we have $(N-s)\sigma^2 \simeq N\sigma^2 \simeq 1\,\text{m}^2$; at the same time, the difference between $\mathbf{r}_i$ and $\mathbf{r}_i + \sigma\hat{\mathbf{n}}$ can be neglected and the volume occupied by the particles ($N\sigma^3 \simeq 10^{-4}\text{cm}^3$) is very small so that the collision between two selected particles is a rather rare event. In this spirit, the Boltzmann-Grad limit has been suggested as a procedure to obtain a closure for Eq. (2.57): $N \to \infty$ and $\sigma \to 0$ in such a way that $N\sigma^2$ remains finite. I stress the fact that (as seen in Sect. 2.1.2) the total number of collisions in the unit of time (for volume and typical velocities both of order 1) is proportional to the total scattering cross section multiplied by N, which for a system of hard spheres gives $N\pi\sigma^2$. The Boltzmann-Grad limit, therefore, states that the single particle collision probability must vanish, but the total number of collisions remains of order 1. Within this limit, the BBGKY hierarchy reads:

$$\frac{\partial P_s}{\partial t} + \sum_{i=1}^{s} \mathbf{v}_i \cdot \frac{\partial P_s}{\partial \mathbf{r}_i} = N\sigma^2 \sum_{i=1}^{s} \int_{\Re^3} \int_{S_+} (P'_{s+1} - P_{s+1})|\mathbf{V}_i \cdot \hat{\mathbf{n}}|d\hat{\mathbf{n}}d\mathbf{v}_* \qquad (2.59)$$

where the arguments of P'_{s+1} and of P_{s+1} are the same as above, except that the position of the $(s+1)-th$ particle ($\mathbf{r}'_*$ and $\mathbf{r}_*$) is equal to $\mathbf{r}_i$ (as $\sigma \to 0$). Equation (2.59) gives a complete description of the time evolution of a Boltzmann gas (i.e. the ideal gas obtained in the Boltzmann-Grad limit), usually called *the Boltzmann hierarchy*.

Finally, the Boltzmann equation is obtained if the *molecular chaos assumption* is taken into account

$$P_2(\mathbf{r}_1, \mathbf{v}_1, \mathbf{r}_2, \mathbf{v}_2, t) = P_1(\mathbf{r}_1, \mathbf{v}_1, t)P_1(\mathbf{r}_2, \mathbf{v}_2, t) \qquad (2.60)$$

for particles that are about to collide (that is when $\mathbf{r}_2 = \mathbf{r}_1 - \sigma\hat{\mathbf{n}}$ and $\mathbf{V}_{12} \cdot \hat{\mathbf{n}} < 0$). This assumption naturally stems from the Boltzmann-Grad limit, as it is reasonable that, in the limit of vanishing single-particle collision rate, two colliding particles are uncorrelated. The lack of correlation of colliding particles is the essence of the molecular chaos assumption. I underline that nothing is said about correlation of particles that have just collided.

With the assumption (2.60) one can rewrite the first equation of the hierarchy (2.59), omitting the 1 subscript (and obvious time dependence) for simplicity:

$$\frac{\partial P(\mathbf{r}, \mathbf{v})}{\partial t} + \mathbf{v} \cdot \frac{\partial P(\mathbf{r}, \mathbf{v})}{\partial \mathbf{r}} = N\sigma^2 \int_{\Re^3} \int_{S_+} (P(\mathbf{r}, \mathbf{v}')P(\mathbf{r}, \mathbf{v}'_*) - P(\mathbf{r}, \mathbf{v})P(\mathbf{r}, \mathbf{v}_*))|\mathbf{V} \cdot \hat{\mathbf{n}}|d\mathbf{v}_* d\hat{\mathbf{n}}$$

$$(2.61)$$

with $\mathbf{v}' = \mathbf{v} - \hat{\mathbf{n}}(\mathbf{V} \cdot \hat{\mathbf{n}})$, $\mathbf{v}'_* = \mathbf{v}_* + \hat{\mathbf{n}}(\mathbf{V} \cdot \hat{\mathbf{n}})$, $\mathbf{V} = \mathbf{v} - \mathbf{v}_*$. This represents the Boltzmann equation for hard spheres. I also observe that the integral in Eq. (2.61) is extended to the hemisphere S_+ but could be equivalently extended to the entire sphere S^2 provided a factor $1/2$ is inserted in front of the integral itself, as changing $\hat{\mathbf{n}} \rightarrow -\hat{\mathbf{n}}$ does not change the integrand.

From a rigorous point of view, the molecular chaos has to be assumed and cannot be proved. However, it has been demonstrated that if the Boltzmann hierarchy has a unique solution for data that satisfy for $t = 0$ a generalized form of chaos assumption:

$$P_s(\mathbf{r}_1, \mathbf{v}_1, \ldots, \mathbf{r}_s, \mathbf{v}_s, t) = \prod_{j=1}^{s} P_1(\mathbf{r}_j, \mathbf{v}_j, t) \qquad (2.62)$$

then Eq. (2.62) holds at any time and therefore the Boltzmann equation is fully justified. Otherwise it has also been proved that if Eq. (2.62) is satisfied at $t = 0$ and the Boltzmann equation (2.61) admits a solution for the given initial data, then the Boltzmann hierarchy (2.59) has at least a solution which satisfy (2.62) at any time t [34, 49].

2.2.4 Collision Invariants and H-theorem

The integral appearing in the right-hand side of Eq. (2.61) is usually called collision integral:

$$Q(P, P) = \int_{\Re^3} \int_{S_+} (P'P'_* - PP_*)|\mathbf{V} \cdot \hat{\mathbf{n}}|d\mathbf{v}_* d\hat{\mathbf{n}} \qquad (2.63)$$

where I have used an intuitive contracted notation (the prime or $*$ must be considered applied to the velocity vector in the argument of the function P). In the collision integral, the position $\mathbf{r}$ is the same wherever the function P appears, and therefore it can be considered a parameter of $Q(P, P)$.

Let us have a look to the integral, for a generic function $\Phi(\mathbf{v})$,

$$\int_{\Re^3} Q(P, P)\Phi(\mathbf{v})d\mathbf{v} = \int_{\Re^3} \int_{\Re^3} \int_{S_+} (P'P'_* - PP_*)\Phi(\mathbf{v})|\mathbf{V} \cdot \hat{\mathbf{n}}|d\mathbf{v}_* d\hat{\mathbf{n}}d\mathbf{v} \qquad (2.64)$$

which can be transformed in many alternative forms, using its symmetries. In particular one can exchange primed and unprimed quantities, as well as starred and unstarred quantities. With manipulations of this sort, it is immediate to get the following alternative form of Eq. (2.64):

$$\int_{\Re^3} Q(P, P)\Phi(\mathbf{v})d\mathbf{v} = \frac{1}{8} \int_{\Re^3} \int_{\Re^3} \int_{S_+} (P'P'_* - PP_*)(\Phi + \Phi_* - \Phi' - \Phi'_*)|\mathbf{V} \cdot \hat{\mathbf{n}}|d\mathbf{v}_* d\hat{\mathbf{n}}\Phi(\mathbf{v})d\mathbf{v}$$

(2.65)

From this equation it comes that if

$$\Phi + \Phi_* = \Phi' + \Phi'_*$$

(2.66)

almost everywhere in velocity space, then the integral of Eq. (2.65) is zero independent of the particular function P. Many authors have proved under different assumptions that the most general solution of Eq. (2.66) is given by

$$\Phi(\mathbf{v}) = C_1 + \mathbf{C}_2 \cdot \mathbf{v} + C_3|\mathbf{v}|^2$$

(2.67)

Furtherly, if $\Phi = \log P$, from Eq. (2.65) it follows that

$$\int_{\Re^3} Q(P, P)\Phi(\mathbf{v})d\mathbf{v} = \frac{1}{8} \int_{\Re^3} \int_{\Re^3} \int_{S_+} (P'P'_* - PP_*) \log(PP_*/P'P'_*)|\mathbf{V} \cdot \hat{\mathbf{n}}|d\mathbf{v}_* d\hat{\mathbf{n}}\Phi(\mathbf{v})d\mathbf{v} \le 0$$

(2.68)

which follows from the elementary inequality $(z - y) \log(y/z) \le 0$ if $y, z \in \Re^+$. This becomes an equality if and only if $y = z$, therefore the equality sign holds in Eq. (2.68) if and only if

$$P'P'_* = PP_*.$$

(2.69)

This is equivalent to two important facts. First, $\Phi + \Phi_* = \Phi' + \Phi'_*$ [taking the logarithms of both sides of Eq. (2.69)], so that one can use the result (2.67) obtaining $P = \exp(C_1 + \mathbf{C}_2 \cdot \mathbf{v} + C_3|\mathbf{v}|^2) = C_0 \exp(-\beta|\mathbf{v} - \mathbf{v}_0|^2)$ where I have defined $C_0 = \exp(C_1)$, $\beta = -C_3$ and $\mathbf{v}_0 = \mathbf{C}_2/2\beta$; this function is called Maxwell-Boltzmann distribution or simply Maxwellian. Second, $Q(P, P) \equiv 0$, i.e. the collision integral identically vanishes for the Maxwellian.

Equation (2.68) is a fundamental result of the Boltzmann theory (it is often called Boltzmann Inequality) and can be fully appreciated with the following discussion. I rewrite the Boltzmann Equation (2.61) with a simplified notation:

$$\frac{\partial P}{\partial t} + \mathbf{v} \cdot \frac{\partial P}{\partial \mathbf{r}} = N\sigma^2 Q(P, P).$$

(2.70)

I multiply both sides by $\Phi = \log P$ and integrate with respect to $\mathbf{v}$, obtaining a transport equation for the quantity Φ:

$$\frac{\partial H}{\partial t} + \frac{\partial}{\partial \mathbf{r}} \cdot \mathbf{j}_H = S_H$$

(2.71a)

$$H = \int_{\Re^3} P \log P d\mathbf{v} \tag{2.71b}$$

$$\mathbf{j}_H = \int_{\Re^3} \mathbf{v} P \log P d\mathbf{v} \tag{2.71c}$$

$$S_H = N\sigma^2 \int_{\Re^3} \log P Q(P, P) d\mathbf{v}. \tag{2.71d}$$

Then Eq. (2.68) states that $S_H \leq 0$ and $S_H = 0$ if and only if P is a Maxwellian. For example, if one looks for a space homogeneous solution of the Boltzmann equation, it happens that

$$\frac{\partial H}{\partial t} = S_H \leq 0 \tag{2.72}$$

that is the famous H-Theorem. It simply states that there exists a macroscopic quantity (H in this case) that decreases as the gas evolves in time and eventually goes to zero when (if and only if) the distribution P becomes a Maxwellian. When the homogeneity is not achievable (due to non-homogeneous boundary conditions) rigorous results are more complicated, but one is still tempted to say that the Maxwellian represents the local asymptotic equilibrium, with the spatial dependence carried by the parameters of this distribution function. For a discussion of the meaning of the H-theorem and the long debate about irreversibility and its many paradoxes, see [20].

2.2.5 The Maxwell Molecules

The collisional integral of Boltzmann equation for hard spheres, Eq. (2.63), contains a term $g = |\mathbf{V} \cdot \hat{\mathbf{n}}|$ which multiplies the probabilities of particles entering or coming out from a collision. In general, the collisional integral must contain the differential collision rate $dR/d\Omega$ for particle coming at a certain relative velocity (in modulus g and direction $\hat{\mathbf{n}}$, or equivalently scattering angle χ centered in the solid angle $d\Omega$), which may be expressed in terms of the scattering cross section s [see for example Eq. (2.19)]:

$$\frac{dR}{d\Omega} = gs(g, \chi) P_2(\mathbf{r}, \mathbf{r} + \sigma\hat{\mathbf{n}}, \mathbf{v}_1, \mathbf{v}_2, t) d\mathbf{v}_2. \tag{2.73}$$

I discussed in Sect. 2.1.2 the fact that the scattering cross section depends strongly on the kind of interaction between the molecules of the gas. For power law repulsive interaction potential $U(r) \sim r^{-(a-1)}$, the scattering angle χ depends on the relative energy $g^2/2$ and on the impact parameter b only through the combination $(g^2 b^{a-1})$. This means that there exists a function $\gamma(\chi)$ such that:

$$b = g^{-2/(a-1)} \gamma(\chi) \tag{2.74}$$

and this means that from relation (2.21) one obtains:

$$gs(g, \chi) \sim g^{1-4/(a-1)} \frac{\gamma(\chi)}{\sin \chi} \frac{d\gamma}{d\chi} \tag{2.75}$$

which holds in $d = 3$. The extension to generic dimension of the last equation is:

$$gs(g, \chi) \sim g^{1-2(d-1)/(a-1)} \frac{\gamma^{d-2}}{(\sin \chi)^{d-2}} \frac{d\gamma}{d\chi} \sim g^{1-2(d-1)/(a-1)} \alpha(\cos \chi). \tag{2.76}$$

Therefore, when $a = 1 + 2(d - 1)$ (i.e. $a = 5$ for $d = 3$ and $a = 3$ for $d = 2$) the collision rate $gs(g, \chi)$ *does not depend upon* g. This property defines the so-called Maxwell molecules [19]. Interaction with $a < 1 + 2(d - 1)$ are called soft interactions (e.g. the electrostatic or gravitational interaction). Interactions with $a > 1 + 2(d - 1)$ are called hard interactions. Hard spheres ($a \rightarrow \infty$) belongs to this set of interactions, with $gs(g, \chi) \sim g$. It has been also studied the Very Hard Particles model, which is characterized by $gs(g, \chi) \sim g^2$, which is not attainable with an inverse power potential, as it requires an interaction harder than the hard sphere interaction.

The advantage of Maxwell molecules is that the Boltzmann equation is greatly simplified, as g does not appear in the collision integral. A further simplification of the Boltzmann equation came from Krook and Wu [32], who studied the Boltzmann equation of Maxwell molecules with an isotropic scattering cross-section, i.e $\alpha = const$, often called Krook and Wu model. A very large literature exists for linear and non-linear model-Boltzmann equations [for a review see [19]]. The importance of the Maxwell molecules model is the possibility of obtaining solutions for it: the general method (extended to other model-Boltzmann equations) is to obtain an expansion in orthogonal polynomial where the expansion coefficients are polynomial moments of the solution distribution function. For Maxwell molecules the moments satisfy a recursive system of differential equations that can be solved sequentially. Given an initial distribution, one can solve the problem if the series expansion converges. Bobylev [8] has shown that if one searches for *similarity* solutions [i.e. solutions with scaling form $P(\mathbf{v}, t) \equiv e^{-\alpha t} F(e^{-\alpha t} \mathbf{v})$], then the solution can be found solving a recursive system of algebraic equation. The Maxwell molecules model has been subject of study also in the framework of the kinetic theory of granular gases [2, 4, 9].

2.2.6 The Enskog Correction

The Boltzmann-Grad limit (see Sect. 2.2.3) restricts the validity of the Boltzmann equation to rarefied gases. This conditions is necessary to consider valid the *Molecular Chaos* which states the independence of colliding particles. In principle, in fact, two colliding particles can be correlated due to an intersection of their collisional histories: one simple possibility is that they may have collided some time before or, alternatively, they may have collided with particles that have collided before. Moreover, the spatial extension of particles (i.e. the fact that they are not really pointlike) restricts the possibilities of motion and as a consequence the degree of independence (this is the so called *excluded volume effect*). All these kinds of correlations become relevant when the gas is not in the situation considered by the Boltzmann-Grad limit, that is when the gas is not rarefied but (either moderately or highly) dense.

The first approach to the problem of not rarefied gases was introduced by Enskog [16]: he did not consider the effects of velocity correlations due to common collisional histories, but simply added to the Boltzmann equation an heuristic correction to take into account short range correlations on positions only. In general the two-body probability distribution function can be written in terms of the one-body functions:

$$P_2(\mathbf{r}_1, \mathbf{v}_1, \mathbf{r}_2, \mathbf{v}_2, t) = g_2(\mathbf{r}_1, \mathbf{v}_1, \mathbf{r}_2, \mathbf{v}_2) P_1(\mathbf{r}_1, \mathbf{v}_1) P_1(\mathbf{r}_2, \mathbf{v}_2) \tag{2.77}$$

where g_2 is the pair correlation function. The Molecular Chaos assumption states that before collisions $g_2(\mathbf{r}_1, \mathbf{r}_1 + \sigma \hat{\mathbf{n}}, \mathbf{v}_1, \mathbf{v}_2) \equiv 1$. In the Enskog theory the Molecular Chaos assumption is modified in the following way:

$$P_2(\mathbf{r}_1, \mathbf{v}_1, \mathbf{r}_1 + \sigma \hat{\mathbf{n}}, \mathbf{v}_2, t) = \varXi(\sigma, n(\mathbf{r}_1)) P_1(\mathbf{r}_1, \mathbf{v}_1) P_1(\mathbf{r}_1 + \sigma \hat{\mathbf{n}}, \mathbf{v}_2) \tag{2.78}$$

i.e. g_2 at contact is a function $\varXi(\sigma, n)$ of σ and local density $n(\mathbf{r}_1)$ only, for particles entering or coming out from a collision. The term $\varXi(\sigma, n)$ becomes a multiplicative constant in front of the collisional integral $Q(P, P)$, giving place to the so-called Boltzmann-Enskog equation. Of course, in a general non-homogeneous situation, the density is a spatially and temporally non-uniform quantity which can be described by a macroscopic field: one may assume (as it is in kinetic theory) that this field changes slowly in space-time, so that the Boltzmann-Enskog equation can be locally solved with constant n as it was a Boltzmann equation with an effective total scattering cross section $\varXi(\sigma, n) N \sigma^2$. For elastic hard disks or hard spheres, spatial correlations may be described by the formulas of Carnahan and Starling [14]:

$$\varXi(\sigma, n) = \frac{1 - 7\phi/16}{(1 - \phi)^2} \quad (d = 2) \tag{2.79a}$$

$$\varXi(\sigma, n) = \frac{1 - \phi/2}{(1 - \phi)^3} \quad (d = 3) \tag{2.79b}$$

where ϕ is the solid fraction ($\phi = n\pi\sigma^2/4$ in $d = 2$, $\phi = n\pi\sigma^3/6$ in $d = 3$). This formula is expected to work well with solid fractions below ϕ_c, where a phase transition takes place [1]. The Enskog correction produces, for example, important corrections to the transport coefficients and to the pressure terms in transport equations.

2.3 The Boltzmann Equation for Granular Gases

The binary collision operator $\overline{T}_-(1, 2)$, for inelastic particles, must be changed [44] according to the inelastic collision rules, Eqs. (2.28a, 2.28b) and (2.29a, 2.29b). It must be noted that when $r = 1$ (elastic collisions), the two set of equations coincide, i.e. the direct or inverse collision are identical transformations. This is not true if $r < 1$. Therefore, in the definition of the inverse binary collision operators at the end of Sect. 2.2.1, that is $T_-(1, 2)$ and $\overline{T}_-(1, 2)$, I have put the same operator b_c that appears in the direct binary collision operators $T_+(1, 2)$ and $\overline{T}_+(1, 2)$, while in general it must be used the operator b'_c that replaces velocities with precollisional velocities [using the transformation given in Eqs. (2.29a, 2.29b)]. The adjoint of inverse binary inelastic collision operator (the only one needed in the following) therefore reads:

$$\overline{T}_-(1, 2) = \sigma^2 \int_{\mathbf{V}_{12}\cdot\hat{\mathbf{n}}>0} d\hat{\mathbf{n}}|\mathbf{V}_{12}\cdot\hat{\mathbf{n}}|\left[\frac{1}{r^2}\delta(\mathbf{r}_1 - \mathbf{r}_2 - \sigma\hat{\mathbf{n}})b'_c - \delta(\mathbf{r}_1 - \mathbf{r}_2 + \sigma\hat{\mathbf{n}})\right]$$

(2.80)

Deriving from this the BBGKY hierarchy and putting in the first equation of it the Molecular Chaos assumption, the Boltzmann Equation for granular gases is obtained [30, 44]:

$$\left(\frac{\partial}{\partial t} + L_1^0\right)P(\mathbf{r}_1, \mathbf{v}_1, t) = N\sigma^2 Q(P, P)$$

(2.81)

$$Q(P, P) = \int d\mathbf{v}_2 \int_{\mathbf{V}_{12}\cdot\hat{\mathbf{n}}>0} d\hat{\mathbf{n}}|\mathbf{V}_{12}\cdot\hat{\mathbf{n}}|\left[\frac{1}{r^2}P(\mathbf{r}_1, \mathbf{v}'_1, t)P(\mathbf{r}_1, \mathbf{v}'_2, t) - P(\mathbf{r}_1, \mathbf{v}_1, t)P(\mathbf{r}_1, \mathbf{v}_2, t)\right]$$

(2.82)

where the primed velocities are defined in Eqs. (2.29a, 2.29b). A major difference with respect to the elastic case is the presence of the factor $1/r^2$ in front of the gain collisional term. This term is the main source of unbalance between gain and loss, and is at the basis of the violation of time reversal symmetry and of the H-theorem (see discussion in Sect. 2.3.6).

This equation has been first studied in the spatially homogeneous case (no spatial gradients, $L_1^0 = 0$), with the Enskog correction (i.e. a multiplying factor $\Xi(\sigma, n)$ in front of the collision integral) by Goldshtein and Shapiro [22] and by Ernst and van Noije [43]. The equation in this case reads

$$\frac{\partial}{\partial t}P(\mathbf{v}_1, t) = \Xi(\sigma, n)n\sigma^2 Q(P, P).$$

(2.83)

2.3.1 Average Energy Loss

It is useful to define a rescaled distribution, under the assumption of *spatial homogeneity*:

$$NP(\mathbf{r}, \mathbf{v}, t) = \frac{n}{v_T^3} \tilde{f}(\mathbf{v}/v_T) \tag{2.84}$$

with (assuming $k_B = 1$) $T(t) = m\langle \mathbf{v}^2\rangle/3 = \frac{1}{3}mv_T^2(t)$ e $\mathbf{c} = \mathbf{v}/v_T$ and n the average number density. One sees that $N^2Q \to n^2 v_T^{-2}\tilde{Q}$ where

$$\tilde{Q} = \int d\mathbf{c}_2 \int_+ d\hat{n} |\mathbf{c}_{12} \cdot \hat{n}| \left[\frac{1}{r^2} \tilde{f}(\mathbf{c}_1', t) \tilde{f}(\mathbf{c}_2', t) - \tilde{f}(\mathbf{c}_1) \tilde{f}(\mathbf{c}_2) \right]. \tag{2.85}$$

The main contribution to the time derivative of temperature is given by the effect of inelastic collisions: in homogeneous situations, where collisions reduce the kinetic energy by a quantity proportional to the kinetic energy itself, one expects to find $\dot{T} \propto T$. The rigorous calculations reads

$$\frac{d}{dt} \left(\frac{3}{2} nT \right) \Big|_{coll} = \int d\mathbf{v} \frac{mv^2}{2} \sigma^2 N^2 Q(P, P)$$

$$= \sigma^2 n^2 v_T \frac{mv_T^2}{2} \int d\mathbf{c}_1 c_1^2 \tilde{Q} = -\sigma^2 n^2 v_T T \mu_2 \tag{2.86}$$

with

$$\mu_p = -\int d\mathbf{c}_1 c_1^p \tilde{Q} \tag{2.87}$$

so that

$$\frac{dT}{dt} \Big|_{coll} = -\zeta(t)T \tag{2.88}$$

where

$$\zeta(t) = \frac{2\sqrt{2}}{3} n\sigma^2 \mu_2 \sqrt{\frac{T}{m}}. \tag{2.89}$$

Computation of μ_2, and therefore of ζ, requires the knowledge of $\tilde{f}(c, t)$.

2.3.2 Sonine Polynomials

It is useful to introduce a polynomial expansion which reveals useful in standard kinetic theory as well as in granular kinetic theory: in fact it serves the purpose of describing small corrections to the Maxwellian. Such small corrections appear in homogeneous granular gases, as well as in all (granular or elastic) dilute gases in spatially non-homogeneous situations. The expansion reads:

$$\tilde{f}(\mathbf{c}) = f_{MB}(\mathbf{c}) \left[1 + \sum_{p=1}^{\infty} a_p S_p(c^2) \right] \tag{2.90}$$

with the basic Maxwellian given by

$$f_{MB}(c) = \pi^{-3/2} exp(-c^2). \tag{2.91}$$

The polynomials S_p are said "Sonine" polynomials (they are in fact associated Laguerre polynomials $S_p^{(m)}$ with $m = d/2 - 1$) and constitute a complete set of orthogonal functions:

$$\int d\mathbf{c}\, f_{MB}(c) S_p(c^2) S_{p'}(c^2) = \frac{2(p+1/2)!}{\sqrt{\pi}\, p!} \delta_{pp'} = \mathcal{N}_p \delta_{pp'} \tag{2.92}$$

In granular homogeneous situations one finds good fit by using expression (2.90) stopping the expansion at $p = 2$. In dimension $d = 3$ the first polynomials read

$$S_0(x) = 1 \tag{2.93}$$
$$S_1(x) = -x + 3/2 \tag{2.94}$$
$$S_2(x) = \frac{x^2}{2} - \frac{5x}{2} + \frac{15}{8} \tag{2.95}$$

It is easy to verify that

$$\langle c^2 \rangle = \frac{3}{2}(1 - a_1) \tag{2.96}$$

and

$$\langle c^4 \rangle = \frac{15}{4}(1 + a_2). \tag{2.97}$$

Note also that

$$N \int d\mathbf{v} \frac{mv^2}{2} P(r, v, t) = \frac{mv_T^2}{2} n \int d\mathbf{c}\, c^2 \tilde{f}(\mathbf{c}) = \langle c^2 \rangle \frac{mv_T^2}{2} n \tag{2.98}$$

and

$$N \int d\mathbf{v} \frac{mv^2}{2} P(r, v, t) = n \frac{m\langle v^2 \rangle}{2} = \frac{3}{2} nT = \frac{3}{2} n \frac{mv_T^2}{2} \tag{2.99}$$

so that $\langle c^2 \rangle = 3/2$ and therefore $a_1 = 0$: the first non trivial coefficient is a_2.

Equations for a_2 are found once a model (boundary conditions) is specified. The explicit expression for μ_2 reads

$$\mu_2 = -\int d\mathbf{c}_1 c_1^2 \int d\mathbf{c}_2 \int_+ d\hat{n} |\mathbf{c}_{12} \cdot \hat{n}| \left[\frac{1}{r^2} \tilde{f}(\mathbf{c}_1', t) \tilde{f}(\mathbf{c}_2', t) - f(\mathbf{c}_1) f(\mathbf{c}_2) \right]$$

(2.100)

By using the Sonine expansion truncated at $p = 2$, it is finally obtained

$$\mu_2 = \sqrt{2\pi}(1 - r^2) \left(1 + \frac{3}{16} a_2 + O(a_2^2) \right).$$

(2.101)

2.3.3 The Homogeneous Cooling State

This is the simplest granular regime: it is assumed spatial homogeneity and absence of any energy injection. The system is initialized with some initial non-trivial velocity distribution.

The rescaled distribution implies the appearance of additional contribution to the time-derivative:

$$\frac{\partial N P}{\partial t} = \frac{n}{v_T^3} \frac{\partial \tilde{f}}{\partial t} + \left(-\frac{3n}{v_T^4} \tilde{f} + \frac{n}{v_T^3} \frac{\partial \tilde{f}}{\partial c_1} \frac{\partial c_1}{\partial v_T} \right) \frac{dv_T}{dt}.$$

(2.102)

The following time evolution equation is obtained:

$$\frac{1}{v_T} \frac{\partial \tilde{f}}{\partial t} - \frac{1}{v_T^2} \frac{\partial (\mathbf{c}_1 \tilde{f})}{\partial \mathbf{c}_1} \frac{dv_T}{dt} = \sigma^2 n \tilde{Q}.$$

(2.103)

Recalling the expression for $\dot{T}(t) = -\zeta(t)T(t)$ as well as for $\zeta(t)$, one can see that

$$\frac{1}{v_T^2} \frac{dv_T}{dt} \Big|_{coll} = \frac{1}{2v_T T} \frac{dT}{dt} = -\frac{1}{3} \sigma^2 n \mu_2$$

(2.104)

is time-independent.

It is usually assumed that a scaling function exists $\tilde{f} \to \tilde{f}_{HC}$ with $\frac{\partial \tilde{f}_{HC}}{\partial t} = 0$. If it exists, it must satisfy

$$\frac{\mu_2}{3} \frac{\partial (\mathbf{c}_1 \tilde{f}_{HC})}{\partial \mathbf{c}_1} = \tilde{Q}.$$

(2.105)

This is the kinetic definition of Homogeneous Cooling State.

The solution of the temperature equation reads:

$$T(t) = \frac{T(0)}{(1 + \frac{\zeta(0)t}{2})^2} \tag{2.106}$$

Eq. (2.106) is known as Haff's law [24].

Using the Sonine approximation truncated at the second polynomial one has

$$\zeta(t) = \frac{4\sqrt{\pi}}{3} n\sigma^2 \sqrt{\frac{T(t)}{m}} (1 - r^2) \left(1 + \frac{3}{16} a_2 + O(a_2^2) \right) = \frac{1 - r^2}{3} \omega_c(t) \tag{2.107}$$

with

$$\omega_c = 4\sqrt{\pi} n\sigma^2 \sqrt{\frac{T(t)}{m}} \left(1 + \frac{3}{16} a_2 + O(a_2^2) \right) \tag{2.108}$$

the collision frequency.

After the Haff's law, it is immediate to realize that

$$\omega_c \sim \frac{1}{1 + \zeta(0)t/2} \tag{2.109}$$

which means that the *cumulated number of collisions* goes as $\sim \ln(1 + \zeta(0)t/2)$. This observation suggests to introduce a new time-scale

$$\tau(t) = \tau_0 \ln(1 + \zeta(0)t/2) \tag{2.110}$$

with arbitrary τ_0, getting

$$\frac{\partial}{\partial t} = \frac{\tau_0 \zeta(0)/2}{1 + \zeta(0)t/2} \frac{\partial}{\partial \tau}. \tag{2.111}$$

This is interesting, since it shows that

$$\frac{1}{v_T(t)} \frac{\partial}{\partial t} = \frac{\tau_0 \zeta(0)/2}{v_t(0)} \frac{\partial}{\partial \tau}. \tag{2.112}$$

Finally, with the new time-scale, one has

$$\frac{\partial \tilde{f}}{\partial \tau} + \frac{n\sigma^2 \mu_2}{3} \frac{\partial (\mathbf{c}_1 \tilde{f})}{\partial \mathbf{c}_1} = \sigma^2 n \tilde{Q} \tag{2.113}$$

equivalent to the Boltzmann equation for particles under the effect of a force

$$F = \frac{n\sigma^2 \mu_2 \mathbf{c}}{3} \tag{2.114}$$

which is equivalent to a *positive* viscosity!

All this equivalence makes sense until the state remains homogeneous. I will show in Chap. 3 that the homogeneous cooling state is unstable for large wavelength perturbations.

Ernst and van Noije [43] have given estimates for the tails of the velocity distribution, using an asymptotic method employed by Krook and Wu [32]. This method assumes that for a fast particle the dominant contributions to the collision integral come from collisions with thermal (bulk) particles and that the gain term of the integral can be neglected with respect to the loss term.

The loss term in the Boltzmann equation reads

$$-\int dc_2 \int_+ d\hat{n} |c_{12}\hat{n}| \tilde{f}(c_1)\tilde{f}(c_2) \approx -\pi c_1 \tilde{f}(c_1). \tag{2.115}$$

If $\tilde{f}$ is isotropic, then $\mathbf{c}\frac{d}{d\mathbf{c}}\tilde{f} = c\frac{d}{dc}\tilde{f}$. Then it remains

$$\mu_2\tilde{f} + \frac{1}{3}\mu_2 c\frac{d}{dc}\tilde{f} = -\pi c\tilde{f} \tag{2.116}$$

and for large c one finds

$$\tilde{f} \sim \exp\left(-\frac{3\pi}{\mu_2}c\right). \tag{2.117}$$

It must be recalled that $\mu_2 \sim (1 - r^2)$, which means that this estimate is valid when $c > 1/(1 - r^2)$.

2.3.4 Inelastic Maxwell Molecules

The inelastic version in one dimension of the Boltzmann equation for Maxwell molecules, discussed in Sect. 2.2.5, reads

$$\partial_\tau P(v, \tau) + P(v, \tau) = \beta \int du\, P(u, \tau) P\left(\beta v + (1 - \beta)u, \tau\right) \tag{2.118}$$

where $\beta = 2/(1 + r)$ and the τ counts the number of collisions per particle. It is interesting to remark that Eq. (2.118) is the master equation of the inelastic version of a process introduced by Ulam [6]: at each step an arbitrary pair is selected and the scalar velocities are transformed according to the rule of Eqs. (2.28a, 2.28b). This model has been considered for the first time by Ben-Naim and Krapivsky [4]. They obtained the evolution of the moments of the velocity distributions. Since at large times, $\langle v^n \rangle \sim \exp(-\tau q_n)$, and the decay rates $q_n \neq nq_2/2$ (they depend non-linearly on n), they argued that such a multiscaling behavior prevents the existence of a rescaled asymptotic distribution f such that $P(v, \tau) \rightarrow f(v/v_0(\tau))$ $/v_0(\tau)$, for large τ, where $v_0^2(\tau) = \int v^2 P(v, \tau) dv = E(\tau)$. On the contrary, the "multiscaling" behavior only indicates the fact that the moments of the rescaled distribution $\int x^n f(x) dx = \langle v^n \rangle / v_0^n$ diverge asymptotically for $n \geq 3$, and does not rule out the possibility of the existence of an asymptotic distribution with power law tails. In fact, the Fourier transform of Eq. (2.118)

$$\partial_\tau \hat{P}(k, \tau) + \hat{P}(k, \tau) = \hat{P}[k/(1 - \beta), \tau]\hat{P}[k/\beta, \tau] \tag{2.119}$$

possesses several self-similar solutions of the kind $\hat{P}(k, \tau) = \hat{f}(kv_0(\tau))$, which correspond to the asymptotic rescaled distribution $P(v, \tau) = f(v/v_0(\tau))/v_0(\tau)$. Many of them do not correspond to physically acceptable velocity distributions [4]. The divergence of the higher moments implies a non analytic structure of $\hat{f}$ in $k = 0$, since $\langle v^n \rangle / v_0^n = (-i)^n \frac{d^n}{dk^n} \hat{f}(k)|_{k=0}$, and represents a guide in the selection of the physical solution, which is

$$f(v/v_0(\tau)) = \frac{2}{\pi \left[1 + (v/v_0(\tau))^2\right]^2} \tag{2.120}$$

corresponding to the self-similar solution $\hat{f}(k) = (1 + |k|) \exp(-|k|)$. Notice that (2.120) is a solution of Eq.(2.119) for every $r < 1$, i.e. the asymptotic velocity distribution does not depend on the value of $r < 1$. The discovery of this exact scaling solution [2] paved the way to a long list of papers by different groups, where the problem in more dimensions was tackled and rigorous results for convergence, uniqueness, etc. were obtained [7].

2.3.5 Bulk Driving

The randomly driven granular gas [introduced in [45, 46]] consists of an assembly of N identical hard objects (spheres, disks or rods) of mass m and diameter σ. I put, for simplicity, $k_B = 1$ (the Boltzmann constant). The grains move in a box of volume $V = L^d$ (L is the length of the sides of the box), with periodic boundary conditions, i.e. opposite borders of the box are identified. The mean free path (calculated exactly in Eq. (2.26) for the case of an homogeneous gas of 3D hard spheres with a Maxwellian distribution of velocities) can be roughly estimated as

$$\lambda = \frac{1}{nS} \tag{2.121}$$

where $n = N/V$ is the mean number density and S is the total scattering cross section. I stress the fact that S has the dimensions of a surface in $d = 3$ ($S \sim \sigma^2$), of a line in $d = 2$ ($S \sim \sigma$) and no dimensions in $d = 1$ (this is consistent with the fact that the diameter, in $d = 1$ is irrelevant).

The dynamics of the gas is obtained as the byproduct of two physical phenomena: continuous interaction with the surroundings and inelastic collisions among the grains. The first ingredient is modeled in the shape of a Langevin equation with exact fulfillment of the Einstein relation [see for example [33]], for the evolution of the velocities of the grains in the free time between collisions. The inelastic collisions follow the usual inelastic rule. The equations of motion for a particle i that is not colliding with any other particle, are:

$$m\frac{d}{dt}\mathbf{v}_i(t) = -\gamma_b \mathbf{v}_i(t) + \sqrt{2\gamma_b T_b}\boldsymbol{\eta}_i(t) \tag{2.122a}$$

$$\frac{d}{dt}\mathbf{x}_i(t) = \mathbf{v}_i(t). \tag{2.122b}$$

I call the parameters $\tau_b = m/\gamma_b$ and T_b *characteristic time of the bath* and *temperature of the bath*, respectively. The function $\boldsymbol{\eta}_i(t)$ is a stochastic process with average $\langle\boldsymbol{\eta}_i(t)\rangle = 0$ and correlations $\langle\eta_i^\alpha(t)\eta_j^\beta(t')\rangle = \delta(t-t')\delta_{ij}\delta_{\alpha\beta}$ (α and β being component indexes) i.e. a standard white noise.

In the dynamics of the N particles, as defined in Eqs. (2.122a, 2.122b) and by the inelastic hard core collision rules, the most important parameters are:

- the coefficient of normal restitution r, which determines the degree of inelasticity;
- the ratio $\rho = \tau_b/\tau_c$ between the characteristic time of the bath and the "global" mean free time between collisions.

On the basis of these two parameters, one can define three fundamental limits of the dynamics of our model:

- the elastic limit: $r \to 1^-$;
- the collisionless limit: $\rho \to 0$ ($\tau_c \gg \tau_b$);
- the cooling limit: $\rho \to \infty$ ($\tau_c \ll \tau_b$).

The *elastic limit* is smooth in dimensions $d > 1$, so that one can consider it equivalent to put $r = 1$. In this case the collisions mix up the components leaving constant the energy (in the center of mass frame as well in the absolute frame). One can assume that, in this limit, the effect of the collisions is that of homogenizing the positions of the particles and making their velocity distribution relax toward the Maxwellian with temperature $T = \langle v^2\rangle/d = \langle v_x^2\rangle$ [this temperature is equal to the starting kinetic energy, but is modified by the relaxation toward T_b due to the Langevin Eqs. (2.122a, 2.122b)]. In one dimension this mixing effect (toward a Maxwellian) is no more at work, as the elastic collisions exactly conserve the starting velocity distribution (the collisions can be viewed as exchanges of labels and the particles as non-interacting walkers).

In the *collisionless limit* we have $\tau_c \gg \tau_b$ and, therefore, the collisions are very rare events with respect to the characteristic time of the bath. In this case we can consider the model as an ensemble of non-interacting Brownian walkers, each following the Eqs. (2.122a, 2.122b). Therefore, whatever r is and in any dimension, the distribution of velocities relaxes in a time τ_b toward a Maxwellian with temperature $T = \langle v^2 \rangle / d = T_b$ with a homogeneous density.

Finally, in the *cooling limit*, the collisions are almost the only events that act on the distribution of velocities, while between collisions the particles move almost ballistically. In this limit (if $r < 1$), the gas can be considered stationary only on observation times very long with respect to the time of the bath τ_b, where the effect of the external driving (the Langevin equation) emerges. For observation times larger than the mean free time τ_c but shorter than τ_b, the gas appears as a *cooling granular gas*.

To conclude this brief discussion on the expected behavior of the randomly driven granular gas model, I sketch a scenario with the presence of two fundamental stationary regimes:

- the "collisionless" stationary regime: when $\rho \ll 1$, i.e. approaching the *collisionless* limit; in this regime one expects, after a transient time of the order of τ_b, the stationary statistics of an ensemble of non-interacting Brownian particles (homogeneous density and Maxwell distribution of velocities, absence of correlations);
- the "colliding" stationary regime: when $\rho \gg 1$, i.e. approaching the *cooling* limit, but observing the system on times larger than τ_b; here, we expect to see anomalous statistical properties.

For this model, the Boltzmann equation includes two additional contributions which are equivalent to the "Fokker-Planck" operators which evolve the velocity distribution in a Langevin equation. The equation therefore reads:

$$\frac{\partial P}{\partial t} = n\sigma^2 Q(P, P) + \frac{\gamma_b}{m} \frac{\partial \mathbf{v} P}{\partial \mathbf{v}} + \frac{\gamma_b}{m} \frac{T_b}{m} \nabla_v P, \qquad (2.123)$$

with $Q(P, P)$ defined in Eq. (2.81). Using the definition of rescaled distribution (2.84), and obviously $\dot{v}_T = 0$ (we are in a statistically stationary state), one gets

$$\frac{\partial \tilde{f}}{\partial t} = v_T n\sigma^2 \tilde{Q} + \frac{\gamma_b}{m} \frac{\partial \mathbf{c} \tilde{f}}{\partial \mathbf{c}} + \frac{\gamma_b}{m} \frac{T_b}{2m} \frac{1}{T} \nabla_c \tilde{f}. \qquad (2.124)$$

From the definition, it follows that

$$T = \frac{m}{d} \langle v^2 \rangle \qquad (2.125)$$

and therefore

$$\langle v\dot{v} \rangle = \frac{\dot{T}}{2m} = -\frac{\gamma_b}{m} \langle v^2 \rangle + \frac{\gamma_b}{m} \frac{T_b}{m} - \zeta \frac{T}{2m}. \qquad (2.126)$$

Imposing $\dot{T} = 0$, in the stationary state, we get

$$T - T_b = \zeta \tau_b T \qquad (2.127)$$

which can be (numerically) solved to obtain T (I remind that $\zeta \propto (1 - r^2)T^{1/2}$).
It is worth noting that r e τ_b appear through a factor $(1 - r^2)\tau_b$.

Assuming that at large velocities $\tilde{Q} \sim -\pi c \tilde{f}$, one finds

$$- \pi v_T n \sigma^2 c \tilde{f} + \frac{\gamma_b}{2m} \frac{T_b}{T} \left(\frac{d^2}{dc^2} + \frac{2}{c} \frac{d}{dc} \right) \tilde{f} + \frac{\gamma_b}{m} \left(3 + c \frac{d}{dc} \right) \tilde{f} = 0. \quad (2.128)$$

This has two different "solutions"

- in the limit $\gamma \to 0$ (with $T_b \to \infty$ with finite γT_b), one has $\tilde{f} \sim \exp(-c^{3/2})$ [43]
- when $\gamma > 0$ one apparently finds $\tilde{f} \sim \exp(-c^2)$ but in this case the approximations (in particular having neglected the gain term in the collisional integral) are not guaranteed.

I conclude this description of the bulk-driving model, by mentioning that recent experiments have demonstrated the relevance of this model for real fluidized granular systems [23, 47].

2.3.6 Looking for a "Granular" H-theorem

The H functional, see Eqs. (2.71a–2.71d), is monotonously non-increasing for an evolution dictated by the homogeneous *elastic* Boltzmann equation. When collisions are inelastic, however, monotonicity of H can no more be proven, and indeed numerical simulations demonstrate that it is no more true [3]. It is worth to mention a recent observation [39] which suggests a possible replacement of the Boltzmann H functional in the case of so-called Boltzmann-Fokker-Planck model (BFP). This model is basically the one discussed in Sect. 2.3.5, precisely it is represented by Eq. (2.123). Variants have also been considered, where the velocities are discretized and the Fokker-Planck operator is replaced by a stochatic jump operator with transition rates that satisfy detailed balance with respect to an equilibrium steady distribution.

The candidate Lyapunov functional is the following

$$H_C(t) = \int d\mathbf{v} P(\mathbf{v}, t) \log \frac{P(\mathbf{v}, t)}{\Pi(\mathbf{v})}, \qquad (2.129)$$

where $\Pi(\mathbf{v})$ is the stationary velocity distribution reached asymptotically. Numerical observations and some analytical arguments indicate that for the BFP model the following relation holds

$$\frac{d H_C(t)}{dt} \leq 0. \qquad (2.130)$$

In particular, in the *elastic* version of the BFP model, the result (2.130) can be demonstrated. Note that the elastic BFP model has a trivial steady state, but a nontrivial dynamics.

The origin of the apparently exact result (2.130) is still unknown and a general demonstration is awaited.

References

1. Alder, B.J., Wainwright, T.E.: Phase transition in elastic disks. Phys. Rev. **127**, 359 (1962)
2. Baldassarri, A.: Marini Bettolo Marconi, U., Puglisi A.: Influence of correlations on the velocity statistics of scalar granular gases. Europhys. Lett. **58**, 14 (2002)
3. Bena, I., Coppex, F., Droz, M., Visco, P., Trizac, E., van Wijland, F.: Stationary state of a heated granular gas: fate of the usual H-functional. Phys. A **370**, 179 (2006)
4. Ben-Naim, E., Krapivsky, P.L.: Scaling, multiscaling, and nontrivial exponents in inelastic collision processes. Phys. Rev. E **66**, 011309 (2002)
5. Bernu, B., Mazighi, R.: One-dimensional bounce of inelastically colliding marbles on a wall. J. Phys. A: Math. Gen. **23**, 5745 (1990)
6. Blackwell, D., Mauldin, R.D.: Ulam's redistribution of energy problem: collision transformations. Lett. Math. Phys. **10**, 149 (1985)
7. Bobylev, A.V., Cercignani, C., Gamba, I.M.: Generalized kinetic Maxwell type models of granular gases. In: Mathematical Models of Granular Matter. Lecture Notes in Mathematics 1937, vol 23. Springer, Berlin (2008).
8. Bobylev, V.: Exact solutions of the nonlinear Boltzmann equation and the theory of relaxation of a maxwellian gas. Teoret. Mat. Fiz. **60**, 280 (1984)
9. Bobylev, A.V., Carrillo, J.A., Gamba, I.M.: On some properties of kinetic and hydrodynamic equations for inelastic interactions. J. Stat. Phys. **98**, 743 (2000)
10. Brilliantov, N.V., Spahn, F., Hertzsch, J.M., Pöschel, T.: Model for collisions in granular gases. Phys. Rev. E **53**, 5382 (1996)
11. Brilliantov, N.V., Pöschel, T.: Kinetic Theory of Granular Gases. Oxford University Press, Oxford (2004)
12. Campbell, C.S., Brennen, C.E.: Computer simulation of granular shear flows. J. Fluid. Mech. **151**, 167 (1985)
13. Campbell, C.S.: Rapid granular flows. Ann. Rev. Fluid Mech. **22**, 57 (1990)
14. Carnahan, W.F., Starling, K.E.: Equation of state for nonattracting rigid spheres. J. Chem. Phys. **51**, 635 (1969)
15. Cercignani, C., Illner, R., Pulvirenti, M.: The Mathematical Theory of Dilute Gases. Springer, Berlin (1994)
16. Chapman, S., Cowling, T.G.: The Mathematical Theory of Nonuniform Gases. Cambridge University Press, London (1960)
17. Clausius, R.: Ueber die mittlere Länge der Wege, welche bei der Molecularbewegung gasförmiger Körper von den einzelnen Molecülen zurückgelegt werden; nebst einigen anderen Bemerkungen über die mechanische Wärmetheorie. Ann. Phys. **181**, 239 (1858)
18. Ernst, M.H., Dorfman, J.R., Hoegy, W.R., van Leeuwen, J.M.J.: Hard-sphere dynamics and binary-collision operators. Physica **45**, 127 (1969)
19. Ernst, H.: Nonlinear model-Boltzmann equations and exact solutions. Phys. Rep. **78**, 1 (1981)
20. Falcioni, M., Vulpiani, A.: Meccanica Statistica Elementare. Springer-Verlag Italia, (2014).
21. Goldhirsch, I., Zanetti, G.: Clustering instability in dissipative gases. Phys. Rev. Lett. **70**, 1619 (1993)
22. Goldshtein, A., Shapiro, M.: Mechanics of collisional motion of granular materials. Part 1. General hydrodynamic equations. J. Fluid Mech. **282**, 75 (1995)

23. Gradenigo, G., Sarracino, A., Villamaina, D., Puglisi, A.: Non-equilibrium length in granular fluids: from experiment to fluctuating hydrodynamics. Europhys. Lett. **96**, 14004 (2011)
24. Haff, P.K.: Grain flow as a fluid-mechanical phenomenon. J. Fluid Mech. **134**, 401 (1983)
25. Herrmann, H.J.: Simulation of granular media. Physica A **191**, 263 (1992)
26. Hertzsch, J.-M., Spahn, F., Brilliantov, N.V.: On low-velocity collisions of viscoelastic particles. J. Phys. II **5**, 1725 (1995)
27. Hopkins, M.A., Louge, M.Y.: Inelastic microstructure in rapid granular flows of smooth disks. Phys. Fluids A **3**, 47 (1991)
28. Huthmann, M., Zippelius, A.: Dynamics of inelastically colliding rough spheres: relaxation of translational and rotational energy. Phys. Rev. E **56**, 6275 (1997)
29. Brey, Javier: J., Ruiz-Montero, M.J., Cubero, D.: Homogeneous cooling state of a low-density granular flow. Phys. Rev. E **54**, 3664 (1996)
30. Brey, Javier: J., Moreno, F., Dufty, J.W.: Model kinetic equation for low-density granular flow. Phys. Rev. E **54**, 445 (1996)
31. Jenkins, J.T., Richman, M.W.: Kinetic theory for plane shear flows of a dense gas of identical, rough, inelastic, circular disks. Phys. Fluids **28**, 3485 (1985)
32. Krook, M., Wu, T.T.: Formation of maxwellian tails. Phys. Rev. Lett. **36**, 1107 (1976)
33. Kubo, R., Toda, M., Hashitsume, N.: Statistical Physics II. Nonequilibrium Stastical Mechanics. Springer, Berlin (1991)
34. Lanford III, O.: The Evolution of Large Classical Systems, vol. 35, p. 1. Springer, Berlin (1975).
35. Luding, S., Clément, E., Blumen, A., Rajchenbach, J., Duran, J.: Anomalous energy dissipation in molecular dynamics simulations of grains: the "detachment effect". Phys. Rev. E **50**, 4113 (1994)
36. Luding, S., Huthmann, M., McNamara, S., Zippelius, A.: Homogeneous cooling of rough dissipative particles: theory and simulations. Phys. Rev. E **58**, 3416 (1998)
37. Lun, C.K.K., Savage, S.B.: A simple kinetic theory for granular flow of rough, inelastic, spherical particles. J. Appl. Mech. **54**, 47 (1987)
38. Lun, C.K.K.: Kinetic theory for granular flow of dense, slightly inelastic, slightly rough spheres. J. Fluid Mech. **233**, 539 (1991)
39. Marconi, U.M.B., Puglisi, A., Vulpiani, A.: About an H-theorem for systems with non-conservative interactions. J. Stat. Mech. **8**, 2 (2013)
40. McNamara, S., Young, W.R.: Inelastic collapse and clumping in a one-dimensional granular medium. Phys. Fluids A **4**, 496 (1992)
41. McNamara, S., Young, W.R.: Inelastic collapse in two dimensions. Phys. Rev. E **50**, R28 (1994)
42. McNamara, S., Luding, S.: Energy nonequipartition in systems of inelastic, rough spheres. Phys. Rev. E **58**, 2247 (1998)
43. van Noije, T.P.C., Ernst, M.H.: Velocity distributions in homogeneous granular fluids: the free and the heated case. Granular Matter **1**, 57 (1998)
44. van Noije, T.P.C., Ernst, M.H., Brito, R.: Ring kinetic theory for an idealized granular gas. Physica A **251**, 266 (1998)
45. Puglisi, A., Loreto, V., Marconi, U.M.B., Vulpiani, A.: Clustering and non-gaussian behavior in granular matter. Phys. Rev. Lett. **81**, 3848 (1998)
46. Puglisi, A., Loreto, V., Marconi, U.M.B., Vulpiani, A.: Kinetic approach to granular gases. Phys. Rev. E **59**, 5582 (1999)
47. Puglisi, A., Gnoli, A., Gradenigo, G., Sarracino, A., Villamaina, D.: Structure factors in granular experiments with homogeneous fluidization. J. Chem. Phys. **136**, 014704 (2012)
48. Schorghofer, N., Zhou, T.: Inelastic collapse of rotating spheres. Phys. Rev. E **54**, 5511 (1996)
49. Spohn, H.: Boltzmann hierarchy and Boltzmann Equation, vol. 1048, p. 207. Springer, Berlin (1984).
50. Visco, P., van Wijland, F., Trizac, E.: Collisional statistics of the hard-sphere gas. Phys. Rev. E **77**, 041117 (2008)
51. Walton, O.R., Braun, R.L.: Stress calculations for assemblies of inelastic spheres in uniform shear. Acta. Mech. **63**, 73 (1986)
52. Walton, O.R., Braun, R.L.: Viscosity, granular-temperature, and stress calculations for shearing assemblies of inelastic, frictional disks. J. Rheol. **30**, 949 (1986)

Chapter 3
Hydrodynamics: A Sea of Grains

Abstract A granular fluid with typical boundary conditions used in laboratory or in silico, will develop structures and inhomogeneities in space and time. When spatial and temporal gradients are small, slow fields such as density, flow velocity and granular temperature evolve accordingly to the equations of granular hydrodynamics. The main steps to derive and close those equations, starting from granular Boltzmann equation, are described in this chapter. The application of the method to common situations are discussed. The interesting and still debated problem of fluctuations is introduced, in the last part of the chapter.

3.1 Granular Kinetic Theory

Fluids can be in spatially non-homogeneous situations. This can be an effect of non-equilibrium initial conditions (the experimentalist sets up the system far from the final situations, and then observes the system relaxing toward it), or a consequence of forcing boundary conditions which keep the system in a non-equilibrium stationary state. For granular fluids there always exists an *intrinsic* energy "sink" which keeps the system out of equilibrium. One can—eventually—apply an external forcing in order to keep the fluid in a stationary state. An example of homogeneous forcing has been discussed in the previous chapter, Sect. 2.3.5. Here, non-homogeneous situations are addressed. An example—due to non-homogeneous forcing (coming from only one boundary)—is shown in Fig. 3.1. The theory sketched in this chapter is however valid independently of the origin of non-homogeneity, as long as it satisfies the criterion of "small gradients". It will be useful, for example, to describe the departure from homogeneity in the cooling regime, where no external driving is present.

© The Author(s) 2015
A. Puglisi, *Transport and Fluctuations in Granular Fluids*,
SpringerBriefs in Physics, DOI 10.1007/978-3-319-10286-3_3

Fig. 3.1 A sketch of an
experiment where the
granular assembly is driven
by gravity plus a (periodically
or stochastic) vibrating wall

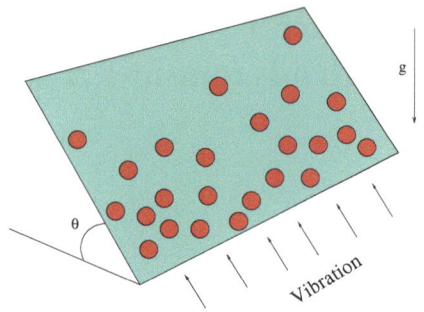

3.1.1 A Sketch of the Chapman-Enskog Approximation Method

The Chapman-Enskog procedure is a way of constructing a non-homogeneous solution, for weak gradients, of the Boltzmann equation [9, 16, 28]. The procedure goes
through a few key steps:

- define densities and fluxes for "slow" variables
- write continuity equations (always valid) for the "slow" quantities
- first assumption: $P(v, r, t)$ depends on r and t only through the above "slow"
  quantities; this means that the Boltzmann equation is replaced by a local Boltzmann
  equation plus equations for the slow parameters
- second assumption: mean free path λ is small with respect to linear size of gradients
  L (which is of order comparable to linear size of the experiment); $\varepsilon = \lambda/L \ll 1$
  is called "Knudsen" number
- for small ε expand fluxes and take only up to linear order in the gradients: "transport
  coefficients" remain to be determined
- for consistency an expansion in powers of ε is set for $P \rightarrow f^{(0)} + \varepsilon f^{(1)} +
  \varepsilon^2 f^{(2)} + \cdots$ and for all spatial and time derivatives: this is equivalent to assume
  that the solution P is the sum of contributions which change on different space
  and time-scales (i.e. different powers of ε)
- these expansions, put into the Boltzmann equation and its supplementary "slow"
  equations, lead to families of equations at different order which can be solved
  separately: each equation governs the evolution of P at a given space-time scale
- at order 0 one has the homogeneous solution (Euler equation for elastic fluids) and
  find $f^{(0)}$
- at order 1 one can find $f^{(1)}$ through its coefficients of the linear expansion in
  gradients; the transport coefficients are functions of these coefficients
- equations at order 2 (hydrodynamics at Navier-Stokes order) are closed now; if
  solved, they can be used to find $f^{(2)}$.

3.1.2 Densities and Fluxes

I assume that a single-particle distribution function is defined, $P(\mathbf{r}, \mathbf{v}, t)$, normalized to give the total number of particles N in the fluid if integrated over the full coordinate-velocities space. P is assumed to be the solution of the Boltzmann Equation (2.62). The particle number density is defined as

$$n(\mathbf{r}, t) = \iiint_\infty d^3 v \, P(\mathbf{r}, \mathbf{v}, t). \tag{3.1}$$

The average molecular velocity is defined as

$$\mathbf{u}(\mathbf{r}, t) = \frac{1}{n(\mathbf{r}, t)} \iiint_\infty d^3 v \, \mathbf{v} \, P(\mathbf{r}, \mathbf{v}, t) \tag{3.2}$$

and this allows to introduce the random velocity vector

$$\mathbf{V}(\mathbf{r}, t) = \mathbf{v} - \mathbf{u}(\mathbf{r}, t) \tag{3.3}$$

which depends on time and position (while $\mathbf{v}$ is independent of t and $\mathbf{r}$) and has zero average:

$$\iiint_\infty d^3 c \, V_i \, P(\mathbf{r}, \mathbf{V}, t) = 0. \tag{3.4}$$

The average fluxes of the molecular quantity $W(\mathbf{v})$ can be expressed as velocity moments of the phase space distribution function:

$$j_W^i(\mathbf{r}, t) = \iiint_\infty d^3 v \, v_i \, W(\mathbf{v}) P(\mathbf{r}, \mathbf{v}, t). \tag{3.5}$$

When $W = m$ one has the mass flux:

$$j_m^i = mn(\mathbf{r}) u_i(\mathbf{r}, t). \tag{3.6}$$

When $W = mv_j$ one has the momentum flux:

$$j_{mv_j}^i = mn(\mathbf{r}, t)\langle v_i v_j \rangle = mn u_i u_j + mn\langle V_i V_j \rangle \tag{3.7}$$

which is a 3×3 symmetric matrix. In the last form two contributions can be recognized, that is the flux due to the bulk (organized) motion and the flux resulting from the random (thermal) motion of the gas particles. This second term is usually called the *pressure tensor* $\mathscr{P}_{ij} = mn\langle V_i V_j \rangle$. One can define, from this discussion, two quantities that are the *scalar pressure* p and the *vector temperature* T_i:

$$p = \frac{1}{3}(\mathscr{P}_{xx} + \mathscr{P}_{yy} + \mathscr{P}_{zz}) \tag{3.8}$$

$$\frac{1}{2}k_B T_i = \frac{1}{2}m\langle V_i^2 \rangle = \frac{1}{2}\frac{\mathscr{P}_{ii}}{n} \tag{3.9}$$

and in the isotropic case $T_i = T$ so that $p = nk_B T$. It can be also defined the stress tensor $\mathscr{T}$ as:

$$\mathscr{T}_{ij} = \delta_{ij}p - \mathscr{P}_{ij} \tag{3.10}$$

which expresses the deviation of the pressure tensor from the equilibrium Maxwellian case (for which $\mathscr{P}_{ij} = p\delta_{ij}$).

Finally, the flux of the quantity $W = mv_j v_k$ is given by:

$$j^i_{mv_j v_k} = mnu_i u_j u_k + u_i \mathscr{P}_{jk} + u_j \mathscr{P}_{ik} + u_k \mathscr{P}_{ij} + \mathscr{Q}_{ijk} \tag{3.11}$$

where $\mathscr{Q}_{ijk} = mn\langle V_i V_j V_k \rangle$ is the generalized heat flow tensor and describes the transport of random energy $V_j V_k$ due to thermal motion V_i of the molecules (for all the permutations of i, j, k).

In Eq. (3.11) three contributions can be recognized: the first term describes the bulk transport of the bulk flux of momentum; the second, third and fourth terms describe a combination of bulk and random momentum fluxes; the last term is the transport of random energy component due to the random motion itself. Often a "classical" heat flow vector is introduced, more intuitive than the generalized heat flow tensor:

$$q_i = \frac{\mathscr{Q}_{ikk}}{2} = n\left\langle V_i \frac{mc^2}{2} \right\rangle. \tag{3.12}$$

3.1.3 Equations for the Densities

Multiplying the Boltzmann equation by 1, v and v^2 and integrating over v, one gets equations for the slow variables:

$$\frac{\partial n}{\partial t} + \nabla \cdot (n\mathbf{u}) = 0 \tag{3.13a}$$

$$\frac{\partial \mathbf{u}}{\partial t} + \mathbf{u} \cdot \nabla \mathbf{u} + (nm)^{-1}\nabla \cdot \mathscr{P} = 0 \tag{3.13b}$$

$$\frac{\partial T}{\partial t} + \mathbf{u} \cdot \nabla T + \frac{2}{3n}\left[\mathscr{P} : (\nabla \mathbf{u}) + \nabla \mathbf{q}\right] + \zeta T = 0. \tag{3.13c}$$

These are the continuity equations and are always valid. The only term which does not appear in the continuity equation for elastic gases is, obviously, the ζT term (indeed $\zeta \equiv 0$ for elatic collisions). I recall that $\zeta(t)$ has been defined in Eq. (2.90) in the previous chapter.

3.1.4 Chapman-Enskog Closure

The Chapman-Enskog procedure consists in

1. change spatial scale $r \rightarrow \varepsilon r$ where $\varepsilon = \lambda/L$, i.e. if old positions were measured in terms of mean free path λ, now the new ones are measured in terms of the characteristic length L which is the macroscopic scale (macroscopic boundary conditions impose spatial variations at this scale); all gradients are transformed as $\nabla \rightarrow \varepsilon \nabla$;
2. for small ε the fluxes can be approximated as linear in the gradients

$$\mathscr{P}_{ij} = p\delta_{ij} - \eta\varepsilon \left(\nabla_i u_j + \nabla_j u_i - \frac{2}{3}\delta_{ij}\nabla \cdot \mathbf{u} \right) \tag{3.14a}$$

$$\mathbf{q} = -\kappa\varepsilon\nabla T - \mu\varepsilon\nabla n \tag{3.14b}$$

the main missing ingredients are, therefore, the coefficients η, κ and μ;
3. the "linear" continuity equations are obtained

$$\frac{\partial n}{\partial t} = -\varepsilon\nabla \cdot (n\mathbf{u}) \tag{3.15a}$$

$$\frac{\partial \mathbf{u}}{\partial t} = -\varepsilon \left(\mathbf{u}\cdot\nabla\mathbf{u} - \frac{1}{nm}\nabla p \right) + \varepsilon^2 \frac{\eta}{mn}\left(\nabla^2\mathbf{u} + \frac{1}{3}\nabla(\nabla\cdot\mathbf{u}) \right) \tag{3.15b}$$

$$\frac{\partial T}{\partial t} = -\zeta T - \varepsilon\left(\mathbf{u}\cdot\nabla T + \frac{2}{3n}p(\nabla\cdot\mathbf{u}) \right) + \varepsilon^2 G, \tag{3.15c}$$

with

$$G = \frac{2\eta}{3n}\left[(\nabla_i u_j)(\nabla_j u_i) + (\nabla_j u_i)(\nabla_j u_i) - \frac{2}{3}(\nabla\cdot\mathbf{u})^2 \right] + \frac{2}{3n}(\kappa\nabla^2 T + \mu\nabla^2 n); \tag{3.16}$$

4. a "normal solution" (also called "Hilbert-class") is assumed for the distribution $P(v, r, t) \rightarrow f[V|n(r, t), u(r, t), T(r, t)]$, (I recall that $V = v - u$), so that derivatives read

$$\frac{\partial f}{\partial t} = \frac{\partial f}{\partial n}\frac{\partial n}{\partial t} + \frac{\partial f}{\partial \mathbf{u}}\cdot\frac{\partial \mathbf{u}}{\partial t} + \frac{\partial f}{\partial T}\frac{\partial T}{\partial t}; \tag{3.17}$$

5. for consistency with the above expansions (and the assumption of "normal" form) one can introduce time-scales which measure the time-variations associated to growing powers of ε (i.e. happening at different spatial scales):

$$\frac{\partial}{\partial t} = \frac{\partial^{(0)}}{\partial t} + \varepsilon \frac{\partial^{(1)}}{\partial t} + \varepsilon^2 \frac{\partial^{(2)}}{\partial t} + \cdots \tag{3.18}$$

where $\frac{\partial^{(i)}}{\partial t}$ stands for a partial derivative with respect to a time which changes on a scale ε^i (a rigorous treatment can be found in [16]);

6. for the same reason, a spatially non-uniform f can be expanded as

$$f = f^{(0)} + \varepsilon f^{(1)} + \varepsilon^2 f^{(2)} + \cdots ; \tag{3.19}$$

7. all these expansions are put into the original Boltzmann equation which (because of the assumed "normal" form) must be supplemented by Eqs. (3.15) for the slow variables; terms at the same order in ε can be solved separately: this must be executed in order of growing powers of ε since at each order the solution at smaller order is needed.

At the smallest (zero) order in ε, the Boltzmann equation with its *supplementary* equations for slow parameters read:

$$\frac{\partial^{(0)} f^{(0)}}{\partial t} = Q(f^{(0)}, f^{(0)}) \tag{3.20a}$$

$$\frac{\partial^{(0)} n}{\partial t} = 0 \tag{3.20b}$$

$$\frac{\partial^{(0)} \mathbf{u}}{\partial t} = 0 \tag{3.20c}$$

$$\frac{\partial^{(0)} T}{\partial t} = -\zeta^{(0)} T. \tag{3.20d}$$

It describes of course a spatially homogeneous situation. The solution of these equations has been already discussed in the previous chapter, it is the Homogeneous Cooling State, i.e. $f^{(0)} = \tilde{f}_{HC}$:

$$f^{(0)} = \frac{n}{v_T^3} \tilde{f}^{(0)} \left(\frac{\mathbf{V}}{v_T} \right) \tag{3.21}$$

with

$$\zeta^{(0)} = -\frac{m}{3nT} \int d\mathbf{v}_1 v_1^2 Q(f^{(0)}, f^{(0)}) = \frac{2}{3} n\sigma^2 \sqrt{\frac{2T}{m}} \mu_2. \tag{3.22}$$

At first order one has:

$$\frac{\partial^{(0)} f^{(1)}}{\partial t} + \left(\frac{\partial^{(1)}}{\partial t} + \mathbf{v} \cdot \nabla \right) f^{(0)} = Q(f^{(0)}, f^{(1)}) + Q(f^{(1)}, f^{(0)}) \qquad (3.23a)$$

$$\frac{\partial^{(1)} n}{\partial t} = -\nabla(n\mathbf{u}) \qquad (3.23b)$$

$$\frac{\partial^{(1)} \mathbf{u}}{\partial t} = -\mathbf{u} \cdot \nabla \mathbf{u} - \frac{1}{nm} \nabla p \qquad (3.23c)$$

$$\frac{\partial^{(1)} T}{\partial t} = -\mathbf{u} \cdot \nabla T - \frac{2}{3} T \nabla \cdot \mathbf{u} - \zeta^{(1)} T. \qquad (3.23d)$$

Putting $f^{(0)} + f^{(1)}$ in the expression for ζ and keeping the first order in ε one has

$$\zeta^{(1)} = 2 \frac{(1 - r^2) m \pi \sigma^2}{24 n T} \int d\mathbf{v}_1 d\mathbf{v}_2 v_{12}^3 f^{(0)} f^{(1)} \qquad (3.24)$$

The above equations are the Euler equations if $r = 1$ (elastic collisions). In elastic case, they describe transport without dissipation (i.e. no viscosity or heat conductivity). Some particular regimes of highly non-homogeneous granular flows may be described by such equations, e.g. the Ideal Granular Hydrodynamics [17].

Knowledge (even formal) of $f^{(0)}$ allows to write an equation for $f^{(1)}$ only. It is necessary to express $\frac{\partial^{(1)} f^{(0)}}{\partial t}$ as

$$\frac{\partial^{(1)} f^{(0)}}{\partial t} = \frac{\partial f^{(0)}}{\partial n} \frac{\partial^{(1)} n}{\partial t} + \frac{\partial f^{(0)}}{\partial \mathbf{u}} \cdot \frac{\partial^{(1)} \mathbf{u}}{\partial t} + \frac{\partial f^{(0)}}{\partial T} \frac{\partial^{(1)} T}{\partial t} \qquad (3.25)$$

The terms in $\frac{\partial^{(1)}}{\partial t}$ are taken from the continuity equations at 1st order. Prefactors are known: $\frac{\partial f^{(0)}}{\partial n} = f^{(0)}/n$, $\frac{\partial f^{(0)}}{\partial \mathbf{u}} = -\frac{\partial f^{(0)}}{\partial \mathbf{V}}$, $\frac{\partial f^{(0)}}{\partial T} = -\frac{1}{2T} \frac{\partial (\mathbf{V} f^{(0)})}{\partial \mathbf{V}}$; analogously one can also write down the "streaming" term $\mathbf{v} \cdot \nabla$, recalling that $p = nT$, getting to

$$\frac{\partial^{(0)} f^{(1)}}{\partial t} + J(f^{(0)}, f^{(1)}) - \zeta^{(1)} T \frac{\partial f^{(0)}}{\partial T} = \mathbf{A} \cdot \nabla \ln T + \mathbf{B} \cdot \nabla \ln n + C_{ij} \nabla_j u_i \qquad (3.26)$$

with $J = -Q(0, 1) - Q(1, 0)$.

R.h.s. depends upon three coefficients which depend only on $f^{(0)}$ and on "slow" fields

$$\mathbf{A} = \mathbf{V}\left[\frac{T}{m}\left(\frac{mV^2}{2T} - 1\right)\frac{1}{V}\frac{\partial}{\partial V} + \frac{3}{2}\right]f^{(0)} \tag{3.27}$$

$$\mathbf{B} = -\mathbf{V}\left(\frac{T}{m}\frac{1}{V}\frac{\partial}{\partial V} + 1\right)f^{(0)} \tag{3.28}$$

$$C_{ij} = \left(V_i V_j - \frac{1}{3}\delta_{ij}V^2\right)\frac{1}{V}\frac{\partial f^{(0)}}{\partial V} \tag{3.29}$$

The most general scalar function depending linearly on vectorial and tensorial gradients is

$$f^{(1)} = \alpha \cdot \nabla \ln T + \beta \cdot \nabla \ln n + \gamma_{ij}\nabla_j u_i \tag{3.30}$$

with coefficients that depend only on V and on space-time through the slow fields.

Putting this form into the Boltzmann equation and comparing terms with same gradients, one obtaines equations for the coefficients of $f^{(1)}$ α, β and γ_{ij}.

The missing transport coefficients can be expressed as functions of the above coefficents of $f^{(1)}$

$$\eta = -\frac{1}{10}\int D_{ij}\gamma_{ji}d\mathbf{V} \tag{3.31a}$$

$$\kappa = -\frac{1}{3T}\int \mathbf{S}\cdot\alpha d\mathbf{V} \tag{3.31b}$$

$$\mu = -\frac{1}{3n}\int \mathbf{S}\cdot\beta d\mathbf{V} \tag{3.31c}$$

$$\zeta^{(1)} = 0, \tag{3.31d}$$

where I have used $\mathbf{S}(V) = \left(mV^2/2 - 5/2T\right)\mathbf{V}$ e $D_{ij} = m\left(V_i V_j - \frac{1}{3}\delta_{ij}V^2\right)$.

In the elastic case $f^{(0)}$ is the Maxwellian f_M, Φ when rescaled to have unitary variance. In this case it is found that $\mathbf{B} = 0$ and therefore $\beta = 0$, leading finally to $\mu = 0$ (Fourier's law).

One gets

$$\eta = -\frac{5}{2\sigma^2}\sqrt{mT/2}\frac{1}{\Omega_\eta[\Phi(c_1), \Phi(c_2)]} \tag{3.32}$$

$$\kappa = -\frac{75}{16\sigma^2}\sqrt{2T/(m)}\frac{1}{\Omega_\kappa[\Phi(c_1), \Phi(c_2)]} \tag{3.33}$$

with the following "pure" numbers

$$\Omega_\eta = \int d\mathbf{c}_1 \int d\mathbf{C}_2 \int d\hat{n} \Theta(-\mathbf{c}_{12} \cdot \hat{n}) |\mathbf{c}_{12} \cdot \hat{n}| \Phi_1(c_1) \Phi_2(c_2)$$
$$\times \left[(\mathbf{c}_1' \cdot \mathbf{c}_2)^2 + (\mathbf{c}_2' \cdot \mathbf{c}_2)^2 - (\mathbf{c}_1 \cdot \mathbf{c}_2)^2 - (\mathbf{c}_2 \cdot \mathbf{c}_2)^2 - \frac{1}{3} c_2^2 \Delta(c_1^2 + c_2^2) \right]$$

(3.34)

and

$$\Omega_\kappa = \int d\mathbf{c}_1 \int d\mathbf{C}_2 \int d\hat{n} \Theta(-\mathbf{c}_{12} \cdot \hat{n}) |\mathbf{c}_{12} \cdot \hat{n}| \Phi_1(c_1) \Phi_2(c_2)$$
$$\times \left(c_2^2 - \frac{5}{2} \right) \left[(\mathbf{c}_1' \cdot \mathbf{c}_2)(c_1')^2 + (\mathbf{c}_2' \cdot \mathbf{c}_2)(c_2')^2 - (\mathbf{c}_1 \cdot \mathbf{c}_2)c_1^2 - (\mathbf{c}_2 \cdot \mathbf{c}_2)c_2^2 \right]$$

(3.35)

obtaining finally

$$\eta = \frac{5}{16\sigma^2} \sqrt{mT/\pi} \tag{3.36a}$$

$$\kappa = \frac{75}{64\sigma^2} \sqrt{T/(m\pi)} \tag{3.36b}$$

$$f^{(0)} + f^{(1)} = f_M(V) \left(1 - \frac{2m\kappa}{5nT^3} \mathbf{S} \cdot \nabla T - \frac{\eta}{nT^2} D_{ij} \nabla_j u_i \right). \tag{3.36c}$$

3.1.5 Inelastic Case

In the inelastic case $f^{(0)}$ is not known analytically, but can be expressed as an expansion in Sonine polynomials, and the coefficients can always be calculated (at any order), for instance stopping at the 2nd order, recalling that $\mathbf{c} = \mathbf{V}/v_T$:

$$f^{(0)} = \left(\frac{n}{v_T^3} \right) \Phi(c)[1 + a_2 S_2(c^2)] \tag{3.37}$$

with

$$S_2(x) = \frac{1}{2} x^2 - \frac{5}{2} x + \frac{15}{8}. \tag{3.38}$$

For consistency, in the coefficients Ω now one must insert $\Omega[(1 + a_2 S_2)\Phi(c_1), \Phi(c_2)]$. One finally gets

$$\eta = \frac{15}{2(1+r)(13-r)\sigma^2} \left(1 + \frac{3}{8}\frac{4-3r}{13-r}a_2\right) \sqrt{mT/\pi} \tag{3.39a}$$

$$\kappa = \frac{75}{2(1+r)(9+7r)\sigma^2} \left(1 + \frac{1}{32}\frac{797+211r}{9+7r}\right) \sqrt{T/(\pi m)} \tag{3.39b}$$

$$\mu = \frac{750(1-r)}{(1+r)(9+7r)(19-3r)n\sigma^2} (1+q(r)a_2) \sqrt{T^3/(\pi m)} \tag{3.39c}$$

$$\zeta = \frac{2}{3}n\sigma^2 \sqrt{\frac{2T}{m}} \mu_2 + \zeta^{(2)}, \tag{3.39d}$$

with $q(r)$ a quite lengthy function of the restitution coefficient r (see [4]). In Eq. (3.39d) we have included the contribution $\zeta^{(2)}$ from second-order gradients to the cooling rate: this contribution is necessary for consistency with the rest of the equations (see below). The contribution, however, for low inelasticities is negligible. A detailed discussion can be found in [4].

It is therefore obtained the solution of the Boltzmann equation at first order in the gradients:

$$f^{(1)}(V) = -\frac{1}{nT^2}\left[\frac{2m}{5T}\mathbf{S}\cdot(\kappa\nabla T + \mu\nabla n) + \eta D_{ij}\nabla_j u_i\right]f_M(V). \tag{3.40}$$

I conclude this section noting that the above procedure (sketched in great generality) leads to a "solution" for the $f^{(i)}(V)$ at order i in the gradients, as well as to closed equations for the slow fields $n(r, t)$, $\mathbf{u}(r, t)$, $T(r, t)$, which include fluxes at order i in the gradients, and therefore (since continuity is given by divergence of fluxes), are at order $i + 1$ in the gradients.

The granular hydrodynamics equations at the Navier-Stokes order, therefore, are Eq. (3.13), with constitutive relations given by Eq. (3.14) and transport coefficients given by (3.39)

3.2 Critiques of Granular Hydrodynamics

In 1995 Du, Li and Kadanoff [15] have published the results of the simulation of a minimal model of granular gas in one dimension. In this model N hard rods (i.e. hard particles in one dimension) move on a segment of length L interacting by instantaneous binary inelastic collisions with a restitution coefficient $r < 1$. To avoid the cooling of the system (due to inelasticity) a thermal wall is placed at one of the boundaries, i.e. when the leftmost particle bounces against the left extreme ($x = 0$) of the segment it is reflected with a new velocity taken out from a Gaussian distribution with variance T. This particle carries the energy to the rest of the system. The main finding of the authors was that even at very small dissipation $1 - r \ll 1$ the profiles predicted by general hydrodynamic equations [26, 30]) were not able to reproduce the essential features of the simulation. In particular the stationary state predicted by hydrodynamics is a flow of heat from the left wall to the right (it goes to zero at the right wall), with no macroscopic velocity flow ($u(x, t) = 0$), a temperature profile $T(x, t)$ which decreases from $x = 0$ to $x = L$, and a density profile inversely proportional to the temperature (as the pressure $p = nT$ is constant throughout the system). The simulations demonstrated that the system settles in an "extraordinary" and non-hydrodynamic state: almost all the particle move slowly and very near the right wall, while kinetic energy is concentrated in the leftmost particle. Reducing the dissipativity $1 - r$ at fixed N the cluster near the wall becomes smaller and smaller. If the heat bath is replaced by a sort of saw-tooth vibrating wall which reflects the leftmost particle always with the same velocity v_0, the evolution of the baricentrum changes in a stationary oscillation very near to the rightmost wall, so that this clustering instability does not disappear. The Boltzmann Equation (see below) can give a qualitative prediction of this clustering phenomena in the limit $N \to \infty$, $1 - r \to 0$ with $N(1 - r) \sim 1$. Further studies [43] have shown that this model has no proper thermodynamic limit, i.e. when $N, L \to \infty$ with $N/L \sim 1$ the mean kinetic energy and the mean dissipated power reduce to zero. This is consistent with the scenario suggested in [15]: the equipartition of energy is broken and the description of the system in terms of macroscopic (slowly varying) quantities no more holds. In this scenario, usual thermodynamic quantities such as mean kinetic energy or mean dissipated power, are not extensive quantities. The thermodynamic limit is recovered if a different setup is considered where the energy is injected "everywhere": this is obtained for instance by coupling each particle to the energy source (e.g. if grains move on a vertically vibrated horizontal plate), as in the model discussed in Sect. 2.3.5 [42].

Kadanoff has addressed, in a general review article [32], a set of experimental situations in which hydrodynamics seems useless, for instance in [29], where a container full of sand is shaken from the bottom and the shaking may be very rapid. The observations indicate that there is a boundary layer of a thickness of few grains near the bottom that is subject to a very rapid dynamics with sudden changes of motion of the particles. At the top of the container, instead, the particles move ballistically encountering very few collisions in their trajectory. Both the top and the bottom of the container cannot be described by hydrodynamics, as the assumption of slow variation of fields or that of scale separation between times (the mean free time must be orders of magnitude lower than the characteristic macroscopic times, e.g. the vibration period) are not satisfied. On the other hand, the slow dynamics regime has been studied, when the vibration is reduced to a rare tapping, so that the system reaches mechanical equilibrium (stop of motion) between successive tappings [34]. The equilibrium is reached at different densities, and—as the tapping is carried on—the "equilibrium" density slowly changes and its evolution depends on many previous instants and not on the very last tap, i.e. is history dependent. This non-locality in time cannot be described by a set of partial differential equations, therefore the hydrodynamic description here fails again. Moreover, in the study of inelastic collapse Kadanoff and Zhou [54] have pointed out that there is a correlation between velocity directions of the particles involved in the collapse: in particular collapse is favored by parallel velocities (because they cannot escape in perpendicular directions). This situation implies a dramatic breakdown of Molecular Chaos assumption and gives evidence of the fact that Inelastic Collapse cannot be described even by a Boltzmann equation.

Along similar lines, Goldhirsch [19–23] raised some points where the hydrodynamics derivation is unclear or possibly ill posed. Using his words, "the notion of a hydrodynamic, or macroscopic description of granular materials is based on unsafe grounds and it requires further study". He addresses two fundamental issues:

1. in granular materials a reference equilibrium state is missing;
2. in granular materials the spatial and temporal scales of the dynamics of the particles are not well separated from the relevant macroscopic scales.

The first problem is more evident than the second. If a molecular gas is left to itself it comes to an equilibrium state given by the stationary solution of the corresponding kinetic equation, e.g. rarefied gases follow the Boltzmann equation. If such an equilibrium state is well defined, perturbations around it can be used as solutions of non-equilibrium problems. Moreover, if external time scales are much larger than the microscopic time scale of relaxation to equilibrium, most of the degrees of freedom of the gas are rapidly averaged and only a few variables are needed for the description of the out of equilibrium dynamics, which obey to macroscopic equations such as Euler or Navier-Stokes equations. If a granular gas is left to itself, instead, the only equilibrium state is an asymptotic death of the motion of all the particles, but before it happens, different kinds of correlations arise leading to strong inhomogeneities (clustering, vortices, shocks, collapse, and so on). In this sense the relaxation to equilibrium has a characteristic time which is infinite and *many other characteristic times* given by different instabilities, due to the non-conservative nature of the collisions. What reference state can be used in a perturbative method like the Chapman-Enskog expansion? In the first derivations of granular hydrodynamics the Maxwell-Boltzmann equilibrium was used [31], in the latest derivations a more rigorous Chapman-Enskog expansion has been followed using solutions of the Enskog-Boltzmann equation by means of a Sonine expansion (which again must be performed around a Maxwell distribution). Goldhirsch has observed however that the limit $(1-r) \to 0$ and $\varepsilon \to 0$ (with ε the Knudsen number, indicating the intensity of the gradients) is smooth and non singular for the granular Boltzmann equation, since the relaxation to local equilibrium takes place in a few collisions per particle, while the effect of (low) inelasticity is relevant on the order of hundreds or thousands of collisions. This means that a perturbative (in $1 - r^2$ and ε) expansion may be applied to the Boltzmann equation around a well suited "elastic" equilibrium, but it is expected to breakdown as $(1 - r)$ or ε are of order ~ 1.

The second issue, raised by Goldhirsch, stems from a more quantitative discussion. He stresses that the lack of scales separation is not only a mere experimental problem: one can in principle think of experiments with an Avogadro number of grains and very large containers. Instead, it is a fundamental problem for granular materials. In fact, such a problem arises not only in granular kinetic theory: when molecular gases are subject to large shear rates or large thermal gradients (i.e. when the velocity field or the temperature field changes significantly over the scale of a mean free path or the time defined by the mean free time) there is no scale separation between the microscopic and macroscopic scales and the gas can be considered mesoscopic. In this case the Burnett and super-Burnett corrections (and perhaps beyond) are of importance and the gas exhibits differences of the normal stress (e.g. $\mathscr{P}_{xx} \neq \mathscr{P}_{yy}$) and other properties characteristic of granular gases. Even if clusters are not expected in molecular gases, strongly sheared gases do exhibit ordering which violates the molecular-chaos assumption. In granular gases this kind of *mesoscopicity* is generic and not limited to strong forcing. Moreover, phenomena like clustering, collapse (and of course avalanches or oscillon excitations) pertain only to granular gases. In mesoscopic systems fluctuations are expected to be stronger and the ensemble averages need not to be representative of their typical values. Furthermore, like in turbulent systems or systems close to second-order phase transitions, in which scale separation vanishes, one expects constitutive relations to be scale dependent, as it happens in granular gases.

The quantitative demonstration of the intrinsic mesoscopic nature of (cooling) granular gases follows from the relation [22]

$$T = C \frac{\gamma^2 \lambda^2}{1 - r^2} \tag{3.41}$$

that relates the local granular temperature with the local shear rate γ and the mean free path λ. The above relation stems from the temperature balance equation in the spatially homogeneous case, by neglecting the (usually small) heat conduction term. It holds until γ can be considered a slow varying (decaying) quantity with respect to the much more rapid decay of the temperature fluctuations (this can be observed by a linear stability analysis and also by the fact that shear modes decay slowly for small wave-numbers—a result of momentum conservation). From the Eq. (3.41) it follows that the ratio between the change of macroscopic velocity over a distance of a mean free path $\lambda\gamma$ and the thermal speed $\sqrt{T}$ is $\sqrt{1 - r^2}/\sqrt{C}$, e.g. $\simeq 0.44$ for $r = 0.9$, that is not small. Thus, except for very low values of $1 - r^2$, the shear rate is always large and the Chapman-Enskog expansion should be carried out beyond the Navier-Stokes order. The above consideration is a simple consequence of the supersonic nature of granular gases [20]. It is clear that a collision between two particles moving in (almost) the same direction reduces the relative velocity, i.e. velocity fluctuations, but not the sum of their momenta, so that in a number of these collisions the magnitude of the velocity fluctuations may become very small with respect to the macroscopic velocities and their differences over the distance of a mean free path. Also the notion of mean free path may become useless: λ is defined as a Galilean invariant, i.e. as the product between the thermal speed $\sqrt{T}$ and the mean free time τ_c; but in a shear experiment the average squared velocity of a particle is given by $\gamma^2 y^2 + T$ (y is the direction of the increasing velocity field), so when $y \gg \sqrt{T}/\gamma$, the distance covered by the particle in the mean free time τ_c is $l(y) = y\lambda\gamma/\sqrt{T} = y\sqrt{1 - r^2}/\sqrt{C}$ and therefore can become much larger that the "equilibrium" mean free path λ and even of the length of the system in the streamwise direction.

Furtherly, the ratio between the mean free time $\tau_c = \lambda/\sqrt{T}$ and the macroscopic characteristic time of the problem $1/\gamma$, using expression (3.41), reads again $\sqrt{1 - r^2}/\sqrt{C}$. This means that also the separation between microscopic and macroscopic time scales is guaranteed only for $r \to 1$. And this result is irrespective of the size of the system or the size of the grains. This lack of separation of time scales poses two serious problem: (a) the fast local equilibration that allows to use local equilibrium as zeroth order distribution function is not obvious; (b) the stability studies are usually performed linearizing hydrodynamic equations, but the characteristic times related to the (stable and unstable) eigenvalues must be of the order of the characteristic "external" time (e.g. $1/\gamma$) which, in this case, is of the order of the mean free time (as just derived), leading to the paradoxical conclusion that the hydrodynamic equations predict instabilities on time scales which they are not supposed to resolve.

Goldhirsch [20] has also shown that the lack of separation of space and time scales leads to scale dependence of fields and fluxes. In particular he has shown that the pressure tensor depends on the scale of the coarse graining used to take space-time averages. This is similar to what happens, for example, in turbulence, where the "eddy viscosity" is scale dependent. Pursuing this analogy, Goldhirsch has noted that an intermittent behavior can be observed in the time series of experimental and numerical measures of the components pressure tensor: single collisions, which are usually averaged over in molecular systems, appear as "intermittent events" in granular systems as they are separated by macroscopic times.

I conclude this section mentioning that the bulk-driving mechanism described in Sect. 2.3.5 is supposed to solve most of the problems discussed above. When every particle interacts with the external bath, the local balances are drastically changed and most of the instabilities are smoothed out. In a stationary state it is expected that the system fluctuates around a well defined "most probable state" (described by a well defined n-particles distribution function, hopefully $n = 1$) and again an expansion around it can be performed. This program has only recently been realized [18].

3.3 Applications of Granular Hydrodynamics

3.3.1 Linear Stability Analysis of the Homogeneous Cooling State

A granular gas prepared with a homogeneous density and no macroscopic flow, at a given temperature $T(0)$, reaches the Homogeneous Cooling State in a few free times τ_c. To study the behavior of small (macroscopic, i.e. for wave vectors of low magnitude $k \ll \min\{2\pi/\lambda, 2\pi/\sigma\}$) fluctuations around this state, a linear stability study of hydrodynamics equations has been performed by several authors (Goldhirsch and Zanetti [22], Deltour and Barrat [14], van Noije et al. [51]). I follow the discussion provided in [50], reviewing their result for the linearized hydrodynamics of rescaled fields. The rescaled fluctuation fields are defined as

$$\delta \tilde{n}(\mathbf{r}, \tau) = \delta n(\mathbf{r}, t)/n, \tag{3.42a}$$

$$\tilde{\mathbf{u}}(\mathbf{r}, \tau) = \mathbf{u}(\mathbf{r}, t)/v_T(t), \tag{3.42b}$$

$$\delta \tilde{T}(\mathbf{r}, \tau) = \delta T(\mathbf{r}, t)/T(t), \tag{3.42c}$$

where I recall $T(t) = \frac{1}{2} m v_T^2(t)$ (see discussion in Sect. 2.3.3). Their Fourier transforms are given by $\delta \tilde{a}(\mathbf{k}, \tau) = \int d\mathbf{r} \exp(-i\mathbf{k} \cdot \mathbf{r}) \delta \tilde{a}(\mathbf{r}, \tau)$, where a is one of $(n, \mathbf{u}, T)$. The vector $\tilde{\mathbf{u}}(\mathbf{k}, \tau)$ can be decomposed in $(d - 1)$ vectors perpendicular to $\mathbf{k}$, called indistinctly $\tilde{\mathbf{u}}_\perp$, and one vector parallel to $\mathbf{k}$, called $\tilde{\mathbf{u}}_\parallel$. The linearized hydrodynamics for these fluctuations is given (in Fourier space) by the following equation:

$$\frac{\partial}{\partial \tau} \delta \tilde{\mathbf{a}}(\mathbf{k}, \tau) = \tilde{\mathscr{M}}(\mathbf{k}) \delta \tilde{\mathbf{a}}(\mathbf{k}, \tau) \tag{3.43}$$

where

$$\tilde{\mathbf{a}} = \begin{cases} (n, u_\perp, u_\parallel, T) \ (d = 2) \\ (n, u_\perp, u'_\perp, u_\parallel, T) \ (d = 3). \end{cases} \tag{3.44}$$

The matrix $\tilde{\mathscr{M}}$ is given (in $d = 2$) by:

$$\tilde{\mathscr{M}} = \begin{pmatrix} 0 & 0 & -ik\lambda & 0 \\ 0 & \gamma_0(1 - k^2 \xi_\perp^2) & 0 & 0 \\ -ik\lambda \left(\frac{1}{2nT\chi_T} \right) & 0 & \gamma_0(1 - k^2 \xi_\parallel^2) & -ik\lambda \left(\frac{p}{2nT} \right) \\ -\gamma_0 g(n) & 0 & -ik\lambda \left(\frac{2p}{dnT} \right) & -\gamma_0(1 + k^2 \xi_T^2) \end{pmatrix} \tag{3.45}$$

with $\gamma_0 = \frac{1-r^2}{2d}$.

Here I have introduced the correlation lengths $\xi_\perp$, $\xi_\parallel$ and ξ_T that depend on the transport coefficients (shear and bulk viscosity and heat conductivity), on the isothermal compressibility $\chi_T = (\partial n/\partial p)_T/n$ and on the pair distribution function $g(n)$ already mentioned. I refer to [50] for detailed calculations of these correlation lengths.

Several facts can be noted in the study of the dispersion relations, i.e. the exponential growth rates of the modes as functions of the wave number [40]. The first is that (in this linear analysis) the evolution of fluctuations of normal velocity components (shear modes, $\tilde{\mathbf{u}}_\perp$) are not coupled with any other fluctuating component. At the same time, all the other components are coupled together. The study of eigenvalues and eigenvectors confirms the fact that the shear modes are not coupled with the other modes. The eigenvectors of the matrix define, beyond the shear modes, three other modes: one heat mode and two sound modes, denoted in the following with the subscripts H and $+$ or $-$ respectively. The associated eigenvalues are $\zeta_\perp(k)$, $\zeta_H(k)$, $\zeta_+(k)$ and $\zeta_-(k)$. It is immediate to see that $\zeta_\perp(k) = \gamma_0(1 - k^2 \xi_\perp^2)$. At low values of k (in the dissipative range defined below) also the heat mode is "pure", as it is given by the longitudinal velocity mode $\tilde{\mathbf{u}}_\|$ only, with eigenvalue $\zeta_H(k) \simeq \gamma_0(1 - \xi_\|^2 k^2)$; in this range the sound modes are combination of density and temperature fluctuations.

The most important result of this analysis is that $\zeta_\perp(k)$ and $\zeta_H(k)$ are *positive* below the threshold values $k_\perp^* = 1/\xi_\perp \sim \sqrt{\gamma_0}$ and $k_H^* \simeq 1/\xi_\| \sim \gamma_0$ respectively, indicating two linearly unstable modes with exponential (in τ) growth rates.

The shear and heat instabilities are well separated at low inelasticity, as $k_\perp^* \sim \sqrt{\gamma_0}$ while $k_H^* \sim \gamma_0$, so that $k_\perp^* \gg k_H^*$. It is also important to note that the linear total size L of the system can suppress the various instability, as the minimum wave number $k_{\min} = 2\pi/L$ can be larger than k_H^* or even than $k_\perp^*$.

Moreover, the study of the eigenvalues of the linear stability matrix, shows that several regimes in the k-space are present:

- for $2\pi/L \ll k \ll \gamma_0/\lambda$ (*dissipative range*), all the eigenvalues are real, so that propagating modes are absent;
- for $\gamma_0/\lambda \ll k \ll \sqrt{\gamma_0}/\lambda$ (*standard range*), the eigenvalues corresponding to sound modes are complex conjugates, so that the sound modes propagate;
- for $\sqrt{\gamma_0}/\lambda \ll k \ll \min\{2\pi/\lambda, 2\pi/\sigma\}$ (*elastic range*) the heat conduction become dominant; in this range the dispersion relations resemble those of an elastic fluid.

The above picture, of course, requires the scale separation $\gamma_0 \ll \sqrt{\gamma_0}$ (valid at low inelasticity).

3.3.2 A Solvable Case: Granular Sedimentation in 2D

An interesting, solvable [4], case is that of hydrodynamics in two dimensions with gravity acting in one direction and a vibrating base. Here $\mathbf{g} = (0, g_e)$ and $g_e < 0$), with the following assumptions: the fields do not depend upon x (the coordinate parallel to the bottom wall), i.e. $\partial/\partial x = 0$, and the system is in a steady state, i.e. $\partial/\partial t = 0$. The continuity equation then reads $\frac{\partial}{\partial y}(n(y)u_y(y)) = 0$ and this can be compatible with the bottom and top walls (where $nv_y = 0$) only if $n(y)v_y(y) = 0$, that is in the absence of macroscopic vertical flow. The equations are written for the dimensionless fields $\tilde{T}(\tilde{y}) = k_B T(y)/(-g_e m\sigma)|_{y=\sigma\tilde{y}}$ and $\tilde{n}(\tilde{y}) = n(y)\sigma^2|_{y=\sigma\tilde{y}}$, while the position y is made dimensionless using $\tilde{y} = y/\sigma$. Finally for the pressure I put $p(y) = \mathscr{P}_{22} = n(y)k_B T(y)$. With the assumption discussed above the equations of granular hydrodynamics read:

$$\frac{d}{d\tilde{y}}(\tilde{n}(\tilde{y})\tilde{T}(\tilde{y})) = -\tilde{n}(\tilde{y}) \tag{3.46}$$

$$\frac{1}{\tilde{n}(\tilde{y})}\frac{d}{d\tilde{y}}Q_r(\tilde{y}) - C(r)\tilde{n}(\tilde{y})\tilde{T}(\tilde{y})^{3/2} = 0 \tag{3.47}$$

where $Q_r(\tilde{y})$ is the granular heat flux expressed by

$$Q_r(\tilde{y}) = A(r)\tilde{T}(\tilde{y})^{1/2}\frac{d}{d\tilde{y}}\tilde{T}(\tilde{y}) + B(r)\frac{\tilde{T}(\tilde{y})^{3/2}}{\tilde{n}(\tilde{y})}\frac{d}{d\tilde{y}}\tilde{n}(\tilde{y}). \qquad (3.48)$$

In the above equations $A(r)$, $B(r)$ and $C(r)$ are dimensionless monotone coefficients, all with the same sign (positive), related to the transport coefficients calculated in Sect. 3.1.5 and explicitly given in [7]. In particular $B(1) = 0$ and $C(1) = 0$, i.e. in the elastic limit there is no dissipation and the heat transport is due only to the temperature gradients, while when $r < 1$ a term dependent upon $\frac{d}{d\tilde{y}}\ln(\tilde{n}(\tilde{y}))$ appears in $Q_r(\tilde{y})$. The use of dimensionless fields eliminates the explicit **g** dependence from the equations, that remains hidden in their structure (the right hand term of equation (3.46), that is due to the gravitational pressure gradient, disappears in the equation for $g = 0$).

A change of coordinate can be applied to Eqs. (3.46) and (3.47) in order to obtain a simpler form:

$$\tilde{y} \rightarrow l(\tilde{y}) = \int_0^{\tilde{y}} \tilde{n}(y')dy' \qquad (3.49)$$

It follows that when y spans the range $[0, L_y]$, the coordinate l spans the range $[0, \sigma/L_x]$. With this change of coordinate it happens that

$$\frac{d}{d\tilde{y}} \rightarrow \tilde{n}(l)\frac{d}{dl} \qquad (3.50)$$

and the first Eq. (3.46) reads:

$$\frac{d}{dl}(\tilde{n}(l)\tilde{T}(l)) = -1 \qquad (3.51)$$

from which is immediate to see that

$$H = \tilde{n}(l)\tilde{T}(l) + l \qquad (3.52)$$

is a constant, i.e. $\frac{d}{dl}H = 0$. This is equivalent to observe that

$$p(y) - g\int_0^y n(y')dy' \qquad (3.53)$$

is constant which is nothing but the Bernoulli theorem for a fluid (at zero velocity) in the gravitational field with the density depending upon the height.

The relation (3.52) is verified by the model simulated in this work for almost all the height of the container, apart of the boundary layer near the bottom driving wall.

Using the coordinate l introduced in (3.49) and the elimination of $\tilde{n}(l)$ using the recognized constant, that is

$$\tilde{n}(l) = \frac{H - l}{\tilde{T}(l)} \tag{3.54}$$

the second Eq. (3.47), after some simplifications, and after a second change of coordinate $l \to s(l) = H - l$, becomes:

$$\frac{\alpha(r)s}{\tilde{T}(s)^{1/2}} \frac{d^2}{ds^2} \tilde{T}(s) - \frac{\alpha(r)s}{2\tilde{T}(s)^{3/2}} \left(\frac{d}{ds}\tilde{T}(s)\right)^2 + \frac{\beta(r)}{\tilde{T}(s)^{1/2}} \frac{d}{ds} \tilde{T}(s) - s\tilde{T}(s)^{1/2} = 0 \tag{3.55}$$

where $\alpha(r) = (A(r) - B(r))/C(r)$, $\beta(r) = (A(r) - \frac{1}{2}B(r))/(C(r))$ are numerically checked to be positive (α is positive for values of r not too low, about $r > 0.3$) and are divergent in the limit $r \to 1$.

The Eq. (3.55) become a linear equation in $\tilde{T}(s)$ as soon as the change of variable $z(s) = \tilde{T}(s)^{1/2}$ is performed:

$$2\alpha(r)s \frac{d^2}{ds^2} z(s) + 2\beta(r) \frac{d}{ds} z(s) - sz(s) = 0 \tag{3.56}$$

giving the solution:

$$z(s) = \mathscr{A} s^{-\nu(r)} I_{\nu(r)}(s/\sqrt{2\alpha}) + \mathscr{B} s^{-\nu(r)} K_{\nu(r)}(s/\sqrt{2\alpha}) \tag{3.57}$$

where I_ν and K_ν are the modified Bessel functions of the first kind and the second kind respectively, $\nu(r) = B(r)/(4(A(r) - B(r)))$ is real and positive for all the values of r greater than the zero of the function $A(r) - B(r)$ (about $r \simeq 0.3$), with $\nu(1) = 0$, while $\mathscr{A}$ and $\mathscr{B}$ are constants that must be determined with assigning the boundary conditions.

Then one can derive the expressions for $\tilde{T}(l)$ and $\tilde{n}(l)$:

$$\tilde{T}(l) = (H - l)^{-2\nu(r)}(\mathscr{A} I_{\nu(r)}((H - l)/\sqrt{2\alpha(r)}) + \mathscr{B} K_{\nu(r)}((H - l)/\sqrt{2\alpha(r)}))^2 \tag{3.58}$$

$$\tilde{n}(l) = \frac{(H - l)^{1+2\nu(r)}}{(\mathscr{A} J_{\nu(r)}((H - l)/\sqrt{2\alpha(r)}) + \mathscr{B} N_{\nu(r)}((H - l)/\sqrt{2\alpha(r)}))^2} \tag{3.59}$$

To calculate the expressions of $\tilde{T}$ and $\tilde{n}$ as a function of the original coordinate $\tilde{y}$ one needs to solve the equation

$$\frac{d}{dl} \tilde{y}(l) = \frac{1}{\tilde{n}(l)} \tag{3.60}$$

putting in it the solution (3.59). However, one can obtain a comparison with the numerical simulations using the new coordinate l. For a discussion of the boundary conditions needed to eliminate the constants H, $\mathscr{A}$ and $\mathscr{B}$ I refer the reader to [7]. In this paper the authors show that the solution fit very well a large region in the bulk but cannot work on the boundary regions near the vibrating bottom and near the open surface. The authors show also that a minimum of the temperature is compatible with the proposed equations. It is important to underline that a minimum in the temperature profile implies an extremal point where the heat conduction term due to temperature gradient vanishes: in such a situation, the only energy transport can be due to the "anomalous" heat conduction due to density gradients [45].

3.3.3 Thermal Convection

Under the effect of gravity, the laterally invariant steady state discussed in Sect. 3.3.2 becomes unstable when the lateral size of the system grows. The most common steady configuration is a convective one, in analogy with thermal convection in molecular fluids. The difference with respect to molecules, is that a granular fluid under gravity spontaneously develops a thermal gradient, i.e. a second external temperature is not required to induce convection.

Granular thermal convection is governed by three dimensionless numbers. One is the Freude number $F_r = \frac{2gL_z}{v_0^2} = \frac{L_z}{z_{\max}}$, where $z_{\max} = v_0^2/2g$ is the maximum height reached by a projectile launched vertically with initial velocity v_0 which is the typical velocity of the vibrating base. The second dimensionless parameter relevant for the problem, the Knudsen number [27]

$$\varepsilon = \frac{2}{\sqrt{\pi}} (\sigma L_z \langle n \rangle)^{-1} = \frac{2}{\sqrt{\pi} N_{layers}} \qquad (3.61)$$

is related to the mean free path and $N_{layers} = N\sigma/L$ is the number of filled layers at rest. Finally, in the case of pure (monodisperse) systems, it is relevant the dissipative parameter

$$R = 8q\varepsilon^{-2} = \pi(1-r)N_{layers}^2 \qquad (3.62)$$

where $q = (1-r)/2$ is a measure of the inelasticity of the system and r is the coefficient of restitution. R depends both on the inelasticity and on the collision rate, since $R \to 0$ if either $r \to 1$ or $N_{layers} \to 0$.

It is useful to recall the hydrodynamic predictions concerning the phase behavior of the system. In [33] it has been presented the following phase diagram (see Fig. 3.2): at fixed F_r and ε, convection rolls appear with increasing inelasticity, i.e. if R overcomes a critical value R_c. Such a value, R_c, is an increasing function of the Knudsen number, ε, which in turn decreases with the number of particles present in the system. With respect to the Froude number, instead R_c is a non-monotonic function of F_r. As shown in Fig. 3.2, at low F_r (i.e. at low gravity or strong shaking) R_c first decreases, i.e. convection is easier to obtain as the gravity increases. R_c however reaches a minimum and then *increases* as the gravity is further increased.

3.3.4 Other Instabilities of Granular Hydrodynamic

The equations of granular hydrodynamics, within linear approximation or by eliminating some terms for special situations, or in its full form, has been applied to explain several laboratory or numerical experiments. A non-exhaustive list of noteworthy phenomena studied by means of granular hydrodynamics includes:

- Gas–liquid or gas–solid phase separation: in this case linear stability analysis is sufficient to predict the transition, while studies of the full non-linear equations can be used to setup a theory analogous to spinodal decomposition [10, 11].
- The Leidenfrost effect can be predicted on the basis of granular hydrodynamics, as a state where density is larger on the top while energy comes from the bottom [38].
- During the pure cooling of granular materials, after the shear instability has broken the HCS, several routes can be followed deep in the non-homogeneous regime. Many hydrodynamic scenario have been proposed, with confirmations coming from event-driven molecular dynamics simulations of hard disks or spheres. In all those scenario, some terms in the transport equations can be neglected. For instance, in one dimension the inviscid Burgers equation, which is hydrodynamics with no pressure in the limit of zero viscosity, gives a rather fair description of the granular shocks observed in simulations [1]. The flow by inertia scenario (where there is no pressure neither viscosity) [39] and ideal granular hydrodynamics [17] (where there are no dissipative flows such as heat and shear transport) are different simplifications which both display the appearance of a singularity at finite

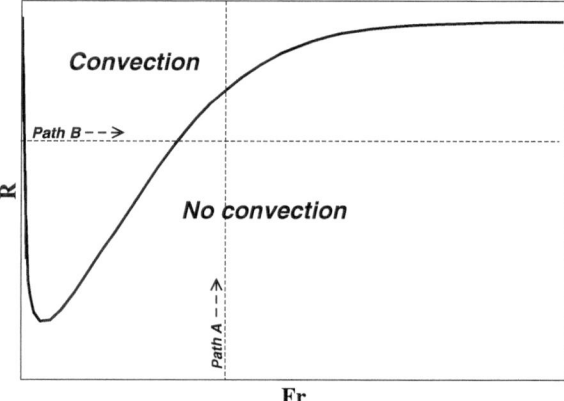

Fig. 3.2 Phase diagram in the plane (F_r, R), at fixes ε, showing the predictions of hydrodynamic theories [27]. Convection is expected increasing R (*path A*), e.g. decreasing the restitution coefficient r, as well as changing F_r in an adequate interval (*path B*). At too low or too high values of F_r (e.g. very low or very high gravity) the system does not reach convection

times. Many of these scenario are inspired by hydrodynamics theories applied in cosmology to describe galaxies and other structures in outer space [44].

- Finally, some theories exist for Non-Newtonian Granular Hydrodynamics, where momentum or heat fluxes are not assumed to be linear in the gradients. These theories do not have the ambition to replace granular hydrodynamics, but to describe and/or enumerate some of the states that can be encountered in granular simulations. A notable example is the so-called Uniform Shear State, which generalizes the simple observation that, due to the energy sink term, the temperature equation can be in a steady state even without heat flow (e.g. with homogeneous temperature and density), but with the only presence of a non-uniform velocity field which generates viscous heating [53].

3.4 Fluctuating Hydrodynamics

A fascinating and largely open problem in granular fluids is that of fluctuations of macroscopic observables. A granular media is often a *small* system, made of a number of grains which can be in the range of 10^2–10^4 grains, that is much smaller than the Avogadro number valid for a molecular fluid. As a consequence, fluctuations in granular fluids are easily observed macroscopically. The problem is made more difficult and interesting by the inherent non-equilibrium nature of granular noise. The relevance of fluctuations represents an interesting point of contact with small molecular systems, e.g. cell sub-units and other systems of biophysical interest, as well as with micro/nano-mechanical devices. Those systems are often in non-equilibrium situations too.

A promising approach is that of fluctuating hydrodynamics, that is the study of fluctuations around an evolution which follows hydrodynamic equations. In summary it consists in separating the evolution of the system into a set of slowly evolving variables and a rapidly relaxing remainder which is treated as noise. This approach is made simpler when the system is at equilibrium: in this case the presence of time-reversal symmetry enforces strong constraints which noise must satisfy. The first example was given by Einstein in his theory of Brownian motion, and appeared in the form of the so-called Einstein relation between diffusivity and mobility. An impressive list of other examples were given in the first decades of the 20th century, culminating with a general theory of linear response, founded around the equilibrium Fluctuation-Dissipation theorem [3]. Fluctuating hydrodynamic near equilibrium [36] makes use of those results through the Green-Kubo relations which relate transport coefficients to the time-correlation of currents or, equivalently, to amplitude of hydrodynamic noise.

Such an elegant and systematic programme fails when considering systems far from equilibrium, such as a granular fluid. Nevertheless, reasonable assumptions for hydrodynamic noise can be found in order to get fluctuating equations. In alternative, a more rigorous derivation from microscopic models can be tried, in few cases with success.

3.4.1 Simple Models of Noise

Here I review a few noteworthy cases where certain properties of the noise are assumed. The validity of such assumptions is checked by deriving results for the amplitude of fluctuations, e.g. correlations, and comparing them with numerical or even experimental results.

For the sake of simplicity, I focus on a particular hydrodynamic mode, that is the shear mode in two dimensions:

$$U_\perp(k, t) = \sum_{j=1}^{N} v_{y,j}(t) e^{-ikx_j(t)}. \tag{3.63}$$

where k is the wave number of chosen mode, and $x_j(t)$, $v_{y,j}(t)$ are the x-coordinate and the y-velocity of particle j at time t, respectively. It was already shown in Sect. 3.3.1 that in granular linearized hydrodynamics (as in standard linearized Navier-Stokes equations) $U_\perp(k, t)$ is decoupled from all other modes.

The main function under investigation, to probe the amplitude of fluctuations, is the rescaled autocorrelation

$$C_\perp(k, t) = \frac{\langle U_\perp(k, 0) U_\perp^*(k, t) \rangle}{2T_y}, \tag{3.64}$$

measured in the steady state, where $T_\beta = \langle v_\beta^2 \rangle$ is the "granular temperature" in the β direction, with β being x or y. Note that $T_y/T_x = 1$ in isotropic systems (e.g. in homogeneous cooling or bulk-driven systems). I also use the shorthand notation $U_\perp(t) \equiv U_\perp(k_{min}, t)$ and $C_\perp(t) \equiv C_\perp(k_{min}, t)$ for the largest mode $k_{min} = 2\pi/L_x$.

At equilibrium (at temperature T), the Landau-Lifshitz fluctuating hydrodynamics, based on Einstein fluctuation formula, predicts

$$\partial_t U_\perp(k, t) = -\nu k^2 U_\perp(k, t) + \xi(k, t), \tag{3.65}$$

where $\nu = \eta/n$ is the kinematic viscosity and $\xi(t)$ is a white Gaussian noise with zero average and

$$\langle \xi(k, t) \xi(k', t') \rangle = \delta_{k', -k} \delta(t - t') 2T N \nu k^2. \tag{3.66}$$

Notice that here I am considering the *extensive* (order $\sim N$) field $U_\perp(k, t)$, and therefore the noise variance appears of order N: the noise associated to the *intensive* field $U_\perp(k, t)/N$ has of course variance scaling with $\sim 1/N$. Based on Eq. (3.65), one has

$$C_\perp(k, t) = \frac{N}{2} e^{-\nu k^2 t}. \tag{3.67}$$

In the inelastic homogeneous cooling regime the Einstein formula for noise amplitude is expected to fail. However, one may insist to apply it in a consistent way (I
will show that such a failure is not dramatic): the trick consists in using the granular
temperature instead of thermostat temperature, which is not defined [49]. Additional
care is required, however, since the amplitude of fluctuations is decaying: indeed,
because of cooling, the kinematic viscosity $\nu(t) \propto \sqrt{T(t)}$ and $T(t)$ decrease with
time (see [5] for details). In particular a constant $q = \nu(t)/\omega_c(t)$ can be defined,
where $\omega_c(t) \propto \sqrt{T(t)}$ is the collision frequency at time t. It is crucial to introduce
a new time-scale τ which is proportional to the cumulative number of collisions [6]:
$d\tau = \omega_c(T(t))dt$. This implies a rescaling of velocity $w(k, \tau) = U_\perp(k, \tau)/\sqrt{T(\tau)}$
and noise $\xi' = \xi/(\omega_c\sqrt{T})$, obtaining the following stochastic equation

$$\partial_\tau w(k, \tau) = -z(k)w(k, t) + \xi'(k, t), \tag{3.68}$$

with $z(k) = -\zeta_\perp(k) = qk^2 - \gamma_0$ which is now time-independent. The statistical
properties of noise are time-independent as well, for the same reason. Indeed

$$\langle \xi'(k, \tau)\xi'(k', \tau') \rangle = \delta_{k',-k}\delta(\tau - \tau')2Nqk^2. \tag{3.69}$$

As already discussed previously (see Sect. 3.3.1) modes are stable only for $z(k) > 0$,
i.e. $k > \sqrt{\frac{\gamma_0\omega_c}{\nu}}$. Those modes reach a steady state. Solving for the corresponding
autocorrelation $C_\perp(k, t)$ in the steady state, one gets

$$C_\perp(k, \tau) = \frac{N}{2} \frac{qk^2}{z(k)} e^{-z(k)t}. \tag{3.70}$$

In the elastic case $\gamma_0 = 0$ and the Einstein-Landau result (3.67) is recovered.

More complicate is the case of a driven granular gas. A first study has been done
in [52], where the hydrodynamics of a gas of inelastic grains which receive energy
by random uncorrelated velocity kicks was considered. In that work hydrodynamic
fluctuations are described using an effective noise that is the sum of an internal and
an external noise. The former is originated from the rapid fluctuations of microscopic
degrees of freedom and its strength can be obtained from an FDR with respect to
internal relaxation, as in the Einstein-Landau picture above. The latter, instead, is
due to the random accelerations received by particles from the external driving. The
strength of this noise is such that, in the steady state, it balances the energy loss due
to the collisions, $\dot{T}|_{coll} = -\zeta T$ (see Eq. 2.90). The result for the shear mode with
small inelasticity is a Langevin equation:

$$\partial_\tau U_\perp(k, t) = -\nu k^2 U_\perp(k, t) + \xi''(k, t), \tag{3.71}$$

with

$$\langle \xi''(k, t)\xi''(k', t') \rangle = \delta_{k',-k}\delta(t - t')NT\left(2\nu k^2 + \zeta\right), \tag{3.72}$$

where the internal ($\propto \nu k^2$) and external ($\propto \zeta$) contributions are recognised. The corresponding autocorrelation $C_\perp(k, t)$ is written as

$$C_\perp(k, t) = \frac{N}{2} \frac{\zeta/2 + \nu k^2}{\nu k^2} e^{-\nu k^2 t}, \tag{3.73}$$

which diverges for small k. The Einstein-Landau result is obtained in the elastic case, $\zeta = 0$.

The situations changes if the external source of energy includes an external viscosity [42], as discussed in Sect. 2.3.5. The fluctuating hydrodynamics for such a model has been recently studied [18, 24] and demonstrated to fairly describe the experimental behavior of a quasi-2D system on a horizontal vibrating plate [25, 41], where roughness of the surface is modelled as a kick+dissipation mechanism which on average defines a "bath temperature" T_b and an interaction time $1/\gamma_b$. Repeating the arguments of [52] for this case, one gets

$$\partial_\tau U_\perp(k, t) = -(\nu k^2 + \gamma_b) U_\perp(k, t) + \xi'''(k, t), \tag{3.74}$$

with

$$\langle \xi'''(k, t) \xi'''(k', t') \rangle = \delta_{k', -k} \delta(t - t') 2N \left(\nu k^2 T + \gamma_b T_b \right). \tag{3.75}$$

Eq. (3.75) leads to the interesting result

$$C_\perp(k, t) = \frac{N}{2} \frac{\nu k^2 T + \gamma_b T_b}{\nu k^2 + \gamma_b} e^{-(\nu k^2 + \gamma_b)t}, \tag{3.76}$$

which turns back to (3.67) in the elastic case when the external bath is detached from the system ($\gamma_b = 0$). Equation (3.76) is interesting because the static structure factor $C_\perp(k, 0)$ defines a *finite* correlation length $\lambda_\perp = \sqrt{\nu/\gamma_b}$, a crucial difference with respect to the non-viscous bath, $\gamma_b = 0$. For lengthscales smaller than $\lambda_\perp$ the structure factor settles to the granular temperature T, while for larger lengthscales, it saturates to T_b. In a real experiment, the measure of $C_\perp(k, 0)$ is an effective way to access to T_b, an "external temperature" not easy to be detected with other means.

To conclude, I mention the case of a steady state obtained by *boundary driving*, which corresponds to spatially inhomogeneous steady hydrodynamic fields. In particular, to avoid complex situations, gravity is considered, as in the model discussed in Sect. 3.3.2. In this case there is not a uniform source of energy throughout the system, but the injection is localized at the bottom of the box. One may divide the system in horizontal layers of width L, equal to the total system's width, and height Δ. The choice of Δ is to be smaller than macroscopic typical lengths but larger than the mean free path. Such a scale is expected to exist if a hydrodynamic description is possible. With such a choice, each layer can be considered spatially homogeneous. In the i-th layer (far enough from the bottom and top boundaries), on average, there is a balance between the energy gain coming from adjacent layers and the energy dissipated in collisions. It is therefore tempting to introduce an effective external noise whose amplitude is determined, as in the case of non-viscous bulk driving discussed above, Eq. (3.71), by the energy balance and is proportional to $N_i \zeta_i T_i$, where the subscript i restricts the measure of averages inside the layer. In [13] the validity of this proposal has been verified numerically with fair success.

3.4.2 Deriving the Fluctuations from the Kinetic Equation

The programme of deriving the hydrodynamic noise from the Boltzmann equation has been carried out in a few cases. The programme requires to setup *fluctuating* Boltzmann equations and to carefully analyze the contribution of the second equation in the Boltzmann hierarchy, which is related to $1/N$ fluctuations even in the dilute limit [12, 46].

In the case of the HCS, rigorous treatment from a fluctuating Boltzmann equation demonstrates the validity of Eq. (3.68) for the rescaled shear mode [5], but with a different expression for the noise:

$$\langle \xi'(k, \tau)\xi'(k', \tau')\rangle = \delta_{k', -k} 2Nk^2 G(|\tau - \tau'|), \tag{3.77}$$

with $G(|s|) \neq \delta(s)$ a function which is discussed in details in [5]. As a matter of fact, a careful study of correlations demonstrate the presence of non-white noise. The steady autocorrelation $C_\perp(\tau)$ obtained from the above Langevin equation is not a simple exponential; anyway it has an exponential tail at large times. In particular one obtains

$$C_\perp(k, 0) = \frac{N}{2} \frac{\nu_1 k^2}{\nu k^2 - \zeta/2} \tag{3.78}$$

$$C_\perp(k, \tau) = \frac{N}{2} \frac{(\nu_1 + \nu_2)k^2}{\nu k^2 - \zeta/2} e^{-(\nu k^2 - \zeta/2)\tau} \qquad \tau \to \infty \tag{3.79}$$

where the two new coefficients ν_1 and ν_2 (the latter is usually smaller than the first) are computed in [5]. In the elastic limit $\zeta \to 0$, $\nu_1 \to \nu$ and $\nu_2 \to 0$ and result (3.67) is recovered. In the inelastic case ($\zeta > 0$) of course one must limit the theory to large enough values of k to have $\nu k^2 - \zeta/2 > 0$: modes with smaller wavenumbers are unstable, as usual. That condition obliges to consider systems smaller than a critical size to avoid the instability.

Similar calculations for the bulk-driving mechanism without viscosity have been performed in [48]. In this case, by adding a simplifying but realistic assumption on two-particles correlation functions, the validity of Eq. (3.71) has been demonstrated, with the same form for the (white) noise.

A more general framework to study fluctuations in the hydrodynamic limit has been setup studying models on a lattice [2]: it goes under the name of "Macroscopic

Fluctuation Theory" (MFT). This programme has been developed well outside the scope of granular kinetic theory. Microscopic probabilistic equations of motion have been proposed to mimic typical transport phenomena, e.g. heat or mass conduction among two reservoir. The simplicity of these models is compensated by a transparent and rigorous procedure to derive the hydrodynamic limit. The theory of large deviations [8, 47] is the natural tool to investigate the role of fluctuations in this limit. Within this approach, one gets a large deviation functional for the fluctuations of the macroscopic observable of interest by contraction, that is by finding the extreme of an action functional related to the average hydrodynamic equations. This procedure, when it can be carried out, is able to describe fluctuations beyond the linear—and therefore Gaussian—approximation, which underly the cases discussed in this section up to this point.

MFT has been applied to a one-dimensional granular kinetic model in [37]. The model consists of a lattice sites equipped with an energy field. The dynamics of the model consists in inelastic collisions, i.e. random choice of neighbouring sites (with probability that can depend on the energy itself) and dissipation plus re-distribution of energy. The fluctuating hydrodynamic equation for the energy density reads

$$\partial_t \rho(x, t) = -\partial_x j(x, t) - \nu \rho(x, t), \tag{3.80}$$

with the fluctuating current $j(x, t) = -D[\rho]\partial_x \rho(x, t) + \xi(x, t)$, ν a dissipation coefficient that is function of microscopic parameters, and $\xi(x, t)$ a Gaussian field which is exactly calculated to satisfy $\langle \xi(x, t) \rangle = 0$ and $\langle \xi(x, t)\xi(x', t') \rangle = N^{-1}\sigma[\rho]\delta(x - x')\delta(t - t')$. A steady state can be obtained by coupling the system with energy reservoirs, for instance to the first and last site, with equal or different temperatures. Even with equal temperatures, the steady state for the inelastic system is out of equilibrium, as a balance of energy current and energy dissipation. In [37] the probability of the fluctuations of the total energy dissipation d averaged on a finite time τ in the steady state is computed to be

$$P_\tau(d) \sim e^{\tau NG(d)}, \tag{3.81}$$

with $G(d)$ that satisfies

$$G(d) = -\min_{j(x)} \int_{-1/2}^{1/2} dx \frac{\{\nu j(x) - D[-j'/\nu]j''(x)\}^2}{2\nu^2\sigma[-j'/\nu]}. \tag{3.82}$$

The solution of this variational problem has been discussed in [37] and yields the so-called *optimal* profile to sustain a certain average value of dissipation d. Other discussions in the same spirit have been given in [35].

References

1. Ben-Naim, E., Chent, S.Y., Doolent, G.D., Redner, S.: Shock-like dynamics of inelastic gases. Phys. Rev. Lett. **83**, 4069 (1999)
2. Bertini, L., De Sole, A., Gabrielli, D., Jona-Lasinio, G., Landim, C.: Macroscopic fluctuation theory. arXiv:1404.6466. (2014)
3. Bettolo Marconi, U.M., Puglisi, A., Rondoni, L., Vulpiani, A.: Fluctuation-dissipation: response theory in statistical physics. Phys. Rep. **461**, 111 (2008)
4. Brey, J.J., Dufty, J.W., Kim, C.S., Santos, A.: Hydrodynamics for granular flow at low density. Phys. Rev. E **58**, 4638 (1998)
5. Brey, J.J., Maynar, P., Garcia de Soria, M.I.: Fluctuating hydrodynamics for dilute granular gases. Phys. Rev. E **79**:051305 (2009)
6. Brey, J.J., Ruiz-Montero, M.J., Moreno, F.: Steady-state representation of the homogeneous cooling state of a granular gas. Phys. Rev. E **69**, 051303 (2004)
7. Brey, J.J., Ruiz-Montero, M.J., Moreno, F.: Hydrodynamics of an open vibrated granular system. Phys. Rev. E **63**, 061305 (2001)
8. Cencini, M., Puglisi, A., Vergni, D., Vulpiani, A., Cecconi, F (eds).: Large Deviations in Physics—Lecture Notes in Physics, volume 885. Springer, Berlin (2014)
9. Chapman, S., Cowling, T.G.: The Mathematical Theory of Nonuniform Gases. Cambridge University Press, London (1970)
10. Clerc, M.G. Argentina, M., Soto, R.: van der Waals-like transition in fluidized granular matter. Phys. Rev. Lett. **89**, 044301 (2002)
11. Clerc, M.G., Cordero, P., Dunstan, J., Huff, K., Mujica, N., Risso, D., Varas, G.: Liquid-solid-like transition in quasi-one-dimensional driven granular media. Nat. Phys. **4**, 249 (2008)
12. Cohen, E.G.D., Ernst, M.H.: Nonequilibrium fluctuations in space. J. Stat. Phys. **25**, 153 (1981)
13. Costantini, G., Puglisi, A.: Fluctuating hydrodynamics in a vertically vibrated granular fluid with gravity. Phys. Rev. E **84**, 031307 (2011)
14. Deltour, P., Barrat, J.-L.: Quantitative study of a freely cooling granular medium. J. Phys. I **7**, 137 (1997). France
15. Du, Y., Li, H., Kadanoff, L.P.: Breakdown of hydrodynamics in a one-dimensional system of inelastic particles. Phys. Rev. Lett. **74**, 1268 (1995)
16. Ferziger, J.H., Kaper, G.H.: Mathematical theory of transport processes in Gases. North-Holland, Amsterdam (1972)
17. Fouxon, I., Meerson, B., Assaf, M., Livne, E.: Formation and evolution of density singularities in hydrodynamics of inelastic gases. Phys. Rev. E **75**, 050301(R) (2007)
18. Garzó, V., Chamorro, M.G., Vega Reyes, F.: Transport properties for driven granular fluids in situations close to homogeneous steady states. Phys. Rev. E **87**, 032201 (2013)
19. Goldhirsch, I.: Kinetics and dynamics of rapid granular flows. In Herrmann, H.J., Hovi, J.-P., Luding, S. (eds.) Physics of dry granular media—NATO ASI Series E 350, p. 371. Kluwer Academic Publishers, Dordrecht (1998)
20. Goldhirsch, I.: Scales and kinetics of granular flows. Chaos **9**, 659 (1999)
21. Goldhirsch, I., Tan, M.-L.: The single-particle distribution function for rapid granular shear flows of smooth inelastic disks. Phys. Fluids **8**, 1752 (1996)
22. Goldhirsch, I., Tan, M.-L., Zanetti, G.: A molecular dynamical study of granular fluids I: the unforced granular gas in two dimensions. J. Sci. Comput. **8**, 1 (1993)
23. Goldhirsch, I., Zanetti, G.: Clustering instability in dissipative gases. Phys. Rev. Lett. **70**, 1619 (1993)
24. Gradenigo, G., Sarracino, A., Villamaina, D., Puglisi, A.: Fluctuating hydrodynamics and correlation lengths in a driven granular fluid. J. Stat. Mech. P08017 (2011)
25. Gradenigo, G., Sarracino, A., Villamaina, D., Puglisi, A.: Non-equilibrium length in granular fluids: from experiment to fluctuating hydrodynamics. Europhys. Lett. **96**, 14004 (2011)
26. Haff, P.K.: Grain flow as a fluid-mechanical phenomenon. J. Fluid Mech. **134**, 401 (1983)
27. He, X., Meerson, B., Doolen, G.: Hydrodynamics of thermal granular convection. Phys. Rev. E **65**, 030301(R) (2002)

28. Huang, K.: Statistical Mechanics. Wiley, Hoboken (1988)
29. Jaeger, H.M., Knight, J.B., Liu, C.-H., Nagel, S.R.: What is shaking in the sandbox? MRS Bull. **XX**, 25 (1994)
30. Jenkins, J.T., Richman, M.W.: Kinetic theory for plane shear flows of a dense gas of identical, rough, inelastic, circular disks. Phys. Fluids **28**, 3485 (1985)
31. Jenkins, J.T., Savage, S.B.: A theory for the rapid flow of identical, smooth, nearly elastic, spherical particles. J. Fluid Mech. **130**, 187 (1983)
32. Kadanoff, L.P.: Built upon sand: theoretical ideas inspired by granular flows. Rev. Mod. Phys. **71**, 435 (1999)
33. Khain, E., Meerson, B.: Onset of thermal convection in a horizontal layer of granular gas. Phys. Rev. E **67**, 021306 (2003)
34. Knight, J.B., Fandrich, C.G., Lau, C.N., Jaeger, H.M., Nagel, S.R.: Density relaxation in a vibrated granular material. Phys. Rev. E **51**, 3957 (1995)
35. Lagouge, M., Bodineau, T.: Current large deviations in a driven dissipative model. J. Stat. Phys. **139**, 201 (2010)
36. Landau, L.D., Lifchitz, E.M.: Physique Statistique. Éditions MIR (1967)
37. Prados, A., Lasanta, A., Hurtado, P.I.: Large fluctuations in driven dissipative media. Phys. Rev. Lett. **107**, 140601 (2011)
38. Meerson, B., Pöschel, T., Bromberg, Y.: Close-packed floating clusters: granular hydrodynamics beyond the freezing point? Phys. Rev. Lett. **91**, 024301 (2003)
39. Meerson, B., Puglisi, A.: Towards a continuum theory of clustering in a freely cooling inelastic gas. Europhys. Lett. **70**, 478 (2005)
40. Orza, J.A.G., Brito, R., van Noije, T.P.C., Ernst, M.H.: Patterns and long range correlations in idealized granular flows. Int. J. Mod. Phys. C **8**, 953 (1997)
41. Puglisi, A., Gnoli, A., Gradenigo, G., Sarracino, A., Villamaina, D.: Structure factors in granular experiments with homogeneous fluidization. J. Chem. Phys. **136**, 014704 (2012)
42. Puglisi, A., Loreto, V., Marconi, U.M.B., Petri, A., Vulpiani, A.: Clustering and non-gaussian behavior in granular matter. Phys. Rev. Lett. **81**, 3848 (1998)
43. Puglisi, A., Loreto, V., Marconi, U.M.B., Vulpiani, A.: A kinetic approach to granular gases. Phys. Rev. E **59**, 5582 (1999)
44. Shandarin, S.F., Zeldovich, Y.B.: The large-scale structure of the universe: turbulence, intermittency, structures in a self-gravitating medium. Rev. Mod. Phys. **61**, 185 (1989)
45. Soto, R., Mareschal, M., Risso, D.: Departure from Fourier's Law for Fluidized Granular Media. Phys. Rev. Lett. **83**, 5003 (1999)
46. Spohn, H.: Boltzmann Hierarchy and Boltzmann Equation, vol. 1048, p 207. Springer, Berlin (1984)
47. Touchette, H.: The large deviation approach to statistical mechanics. Phys. Rep. **478**, 1 (2009)
48. Trizac, E., Maynar, P., de Soria, M.I.G.: Fluctuating hydrodynamics for driven granular gases. Eur. Phys. J. **179**, 123 (2009). (special topics)
49. van Noije, T.C.P., Ernst, M.H., Brito, R., Orza, J.A.G.: Mesoscopic theory of granular fluids. Phys. Rev. Lett. **79**, 411 (1997)
50. van Noije, T.P.C., Ernst, M.H.: Cahn-hilliard theory for unstable granular flows. Phys. Rev. E **61**, 1765 (2000)
51. van Noije, T.P.C., Ernst, M.H., Brito, R., Orza, J.A.G.: Mesoscopic theory of granular fluids. Phys. Rev. Lett. **79**, 411 (1997)
52. van Noije, T.P.C., Ernst, M.H., Trizac, E., Pagonabarraga, I.: Randomly driven granular fluids: large-scale structure. Phys. Rev. E **59**, 4326 (1999)
53. Vega Reyes, F., Santos, A., Garzó, V.: Non-newtonian granular hydrodynamics. What do the inelastic simple shear flow and the elastic fourier flow have in common? Phys. Rev. Lett. **104**, 028001 (2010)
54. Zhou, T., Kadanoff, L.P.: Inelastic collapse of three particles. Phys. Rev. E **54**, 623 (1996)

Chapter 4
Tracer's Diffusion: Swimming Through the Grains

Abstract In this chapter I consider the stochastic dynamics of an intruder in a granular fluid. Under the same assumptions used to derive the Boltzmann equation, a Master Equation for the intruder's velocity is derived. In the limit of large intruder's mass, the dynamics is described by an Ornstein-Uhlenbeck process. I discuss the effects of collisions' inelasticity and of non-Gaussian properties of the surrounding gas. When the shape of the intruder breaks some spatial symmetry, part of the energy dissipated in collisions can be converted in useful work. A granular Brownian motor is then realized.

4.1 The Markovian Limit

A starting point to study the dynamics of an intruder in a granular fluid is the kinetic equation for a two-species mixture [33]. Again I focus on the simplest case of inelastic hard core interactions for smooth disks or spheres, with a constant restitution coefficient. The kinetic theory of this model has been extensively described in Chap. 2. In order to set the system in a steady state, I consider an external driving modelled as a thermal bath with temperature T_b and viscosity γ_b, as discussed in Sect. 2.3.5. The two species correspond to the host fluid and the intruder, whose velocities are denoted here as $\mathbf{v}$ and $\mathbf{V}$, respectively. The gas particles have mass m and diameter σ, while the intruder has mass M and diameter Σ. I also define the following quantities which will be useful: $\varepsilon = \sqrt{m/M}$ and $\chi = n(\sigma/2 + \Sigma/2)^{d-1}$ (with n the gas density).

Under the assumption of Molecular Chaos, two coupled Boltzmann equations can be written [5], for the time derivative of the separated velocity distributions $p(\mathbf{v}, t)$ and $P(\mathbf{V}, t)$:

$$\frac{\partial P(\mathbf{V}, t)}{\partial t} = \int d\mathbf{V}'[W_{tr}(\mathbf{V}|\mathbf{V}')P(\mathbf{V}', t) - W_{tr}(\mathbf{V}'|\mathbf{V})P(\mathbf{V}, t)] + \mathcal{B}_{tr}P(\mathbf{V}, t)$$

$$\frac{\partial p(\mathbf{v}, t)}{\partial t} = \int d\mathbf{v}'[W_g(\mathbf{v}|\mathbf{v}')p(\mathbf{v}', t) - W_g(\mathbf{v}'|\mathbf{v})p(\mathbf{v}, t)] + \mathcal{B}_g p(\mathbf{v}, t)$$

$$+ \chi Q[\mathbf{v}|p, p], \tag{4.1}$$

© The Author(s) 2015
A. Puglisi, *Transport and Fluctuations in Granular Fluids*,
SpringerBriefs in Physics, DOI 10.1007/978-3-319-10286-3_4

where it must be immediately said that W_{tr} and W_g depend upon the distributions $P(\mathbf{V}, t)$ and $p(\mathbf{v}, t)$ and therefore do not represent Markovian transition rates, as discussed in details below.

In Eq. (4.1), $Q[\mathbf{v}|p, p]$ is the collision operator for the gas particle-particle inter-actions, which has been discussed in Chap. 2. Collisional *cross* terms are also present, for the tracer and the gas particles, respectively. The related transition rates take the form

$$
\begin{aligned}
W_{tr}(\mathbf{V}|\mathbf{V}') = \chi \int d\mathbf{v}' \int d\hat{\mathbf{n}}\, p(\mathbf{v}', t) \Theta\left[-(\mathbf{V}' - \mathbf{v}') \cdot \hat{\mathbf{n}}\right] (\mathbf{V}' - \mathbf{v}') \cdot \hat{\mathbf{n}} \\
\times \delta^{(d)}\left\{\mathbf{V} - \mathbf{V}' + k(\varepsilon, r)\left[(\mathbf{V}' - \mathbf{v}') \cdot \hat{\mathbf{n}}\right]\hat{\mathbf{n}}\right\},
\end{aligned}
\tag{4.2}
$$

with $k(\varepsilon, r) = \frac{\varepsilon^2}{1+\varepsilon^2}(1 + r)$, and

$$
\begin{aligned}
W_g(\mathbf{v}|\mathbf{v}') = \frac{\chi}{N} \int d\mathbf{V}' \int d\hat{\mathbf{n}}\, P(\mathbf{V}', t) \Theta\left[-(\mathbf{V}' - \mathbf{v}') \cdot \hat{\mathbf{n}}\right] (\mathbf{V}' - \mathbf{v}') \cdot \hat{\mathbf{n}} \\
\times \delta^{(d)}\left\{\mathbf{v} - \mathbf{v}' + \frac{1+r}{1+\varepsilon^2}\left[(\mathbf{v}' - \mathbf{V}') \cdot \hat{\mathbf{n}}\right]\hat{\mathbf{n}}\right\},
\end{aligned}
\tag{4.3}
$$

where $\Theta(x)$ is the Heaviside step function and $\delta^{(d)}(x)$ is the Dirac delta function in d dimensions.

Finally, the operators $\mathscr{B}_{tr}$ and $\mathscr{B}_g$ take into account the interactions with the thermal bath, as discussed in Sect. 2.3.5:

$$
\mathscr{B}_{tr} P(\mathbf{V}, t) = \frac{\gamma_b}{M} \frac{\partial}{\partial \mathbf{V}} [\mathbf{V} P(\mathbf{V}, t)] + \frac{\gamma_b T_b}{M} \Delta_V [P(\mathbf{V}, t)]
\tag{4.4}
$$

$$
\mathscr{B}_g p(\mathbf{v}, t) = \frac{\gamma_b}{m} \frac{\partial}{\partial \mathbf{v}} [\mathbf{v} p(\mathbf{v}, t)] + \frac{\gamma_b T_b}{m} \Delta_v [p(\mathbf{v}, t)],
\tag{4.5}
$$

where Δ_v is the Laplacian operator with respect to the velocity.

It is important to underline that Eqs. (4.1) do not describe a Markovian process. Indeed the transition rates at a given time depend on the probabilities themselves. A second important observation concerns energy *equipartition* between the two species. It has been shown that when $r < 1$ the two species do not achieve equipartition, that is $m \langle v^2 \rangle \neq M \langle V^2 \rangle$ [11, 23]. In the following I will use T to denote the granular temperature of the gas and T_{tr} to denote the granular temperature of the intruder.

4.1.1 Decoupling the Gas from the Tracer

The system of Boltzmann equations (4.1) is simplified when the quantities $P(\mathbf{V}, t)$ and $p(\mathbf{v}, t)$ significantly change on well-separated characteristic time scales: this situation is achieved when $\chi/N \ll 1$, so that $W_g \sim 0$. Then one may safely assume that the probability distribution function $p(\mathbf{v})$ is stationary. The assumption of stationary $p(\mathbf{v})$ implies that the first equation of the mixture (the evolution of the intruder probability) is *linear* in $P(\mathbf{V})$: it becomes a Master Equation for a Markov process with transition rate W_{tr}.

A first approximation, often not far from numerical or experimental evidence, is to take the steady $p(\mathbf{v})$ to be a Gaussian function with variance T/m:

$$p(\mathbf{v}) = \frac{1}{(2\pi T/m)^{d/2}} \exp\left[-\frac{m\mathbf{v}^2}{2T}\right]. \tag{4.6}$$

When necessary, this approximation can be improved by including its first Sonine non-trivial correction (the second polynomial). The discussion of Sonine corrections has been carried out in Sect. 2.3.2.

4.1.2 The Transition Rate

When $p(\mathbf{v})$ is stationary, one can calculate the transition rate for the intruder.

I first discuss in detail what happens during a collision in order to understand the physical meaning of the final expression. Then I give a rigorous derivation of the transition rate. The collision rule reads

$$\mathbf{V}' = \mathbf{V} - k(\varepsilon, r)[(\mathbf{V} - \mathbf{v}) \cdot \hat{\mathbf{n}}]\hat{\mathbf{n}} \tag{4.7}$$

where $\hat{\mathbf{n}}$ is the direction joining the centers of the two colliding particles. There are some consequences of the collision rules which have to be remarked. For simplicity I assume to be in dimension $d = 2$.

- $\Delta\mathbf{V} = \mathbf{V}' - \mathbf{V}$ is parallel to $\hat{\mathbf{n}}$ with $\hat{n}_x = \cos\theta$ and $\hat{n}_y = \sin\theta$ and $\theta = \arctan\frac{\Delta V_y}{\Delta V_x}$. The fact that $(\mathbf{V} - \mathbf{v}) \cdot \hat{\mathbf{n}}$ must be negative determines completely the angle θ, i.e. the unitary vector $\hat{\mathbf{n}}$. From here on, I call $\Delta V \equiv \Delta V_n \equiv \Delta\mathbf{V} \cdot \hat{\mathbf{n}}$;
- from (4.7) one has $v_n \equiv \mathbf{v} \cdot \hat{\mathbf{n}} = \frac{\Delta V}{k(\varepsilon, r)} + V_n$;
- the component of $\mathbf{v}$ which is not determined by V and ΔV is the one orthogonal to $\hat{\mathbf{n}}$. I call $\hat{\tau}$ the direction perpendicular to $\hat{\mathbf{n}}$, i.e. the vector of component $(-\sin\theta, \cos\theta)$. I define $v_\tau = \mathbf{v} \cdot \hat{\tau}$.

From the above discussion, it follows that the transition probability for the intruder to change velocity during a collision, going from $\mathbf{V}$ to $\mathbf{V}'$, must be

$$W_{tr}(\mathbf{V}'|\mathbf{V}) = C(\mathbf{V}, \mathbf{V}') \int d v_\tau \, p(\mathbf{v}) \tag{4.8a}$$

$$\mathbf{v} = v_n \hat{\mathbf{n}} + v_\tau \hat{\tau} \tag{4.8b}$$

$$\hat{\mathbf{n}} = (\cos\theta, \sin\theta) \tag{4.8c}$$

$$\hat{\tau} = (-\sin\theta, \cos\theta) \tag{4.8d}$$

$$\theta = \arctan\frac{\Delta V_y}{\Delta V_x} \tag{4.8e}$$

$$v_n = \frac{\Delta V}{k(\varepsilon, r)} + V_n. \tag{4.8f}$$

The function C does not depend on the host gas $p(\mathbf{v})$ and must be of dimensions $1/length$. Therefore W_{tr} has dimensions $1/(velocity^d time)$ which is expected because W_{tr} is a rate of change of the velocity pdf (in d dimensions).

Now, I want to obtain the complete result, starting from the expression (4.2) for the intruder's transition rate. Using that for a generic d-dimensional vector $\mathbf{r} = r\hat{\mathbf{r}}$ one has $\delta(\mathbf{r} - \mathbf{r}_0) = \frac{1}{r_0^{d-1}}\delta(r - r_0)\delta(\hat{\mathbf{r}} - \hat{\mathbf{r}}_0)$, one may rewrite (4.2) as

$$W_{tr}(\mathbf{V}'|\mathbf{V}) = \chi \int d\mathbf{v} \int d\hat{\omega} \Theta[(\mathbf{V} - \mathbf{v}) \cdot \hat{\omega}] \frac{|(\mathbf{V} - \mathbf{v}) \cdot \hat{\omega}|}{\Delta v^{d-1}} p(\mathbf{v})\delta(\hat{\mathbf{n}} + \hat{\omega})\delta\left(\Delta V + k(\varepsilon, r)|(\mathbf{V} - \mathbf{v}) \cdot \hat{\omega}|\right). \tag{4.9}$$

Then, performing the angular integration over $\hat{\omega}$, one obtains:

$$W_{tr}(\mathbf{V}'|\mathbf{V}) = \chi \int d\mathbf{v}\Theta[(\mathbf{V} - \mathbf{v}) \cdot \hat{\omega}] \frac{|(\mathbf{V} - \mathbf{v}) \cdot \hat{\omega}|}{\Delta v^{d-1}} p(\mathbf{v})\delta\left(\Delta V + k(\varepsilon, r)|(\mathbf{V} - \mathbf{v}) \cdot \hat{\mathbf{n}}|\right). \tag{4.10}$$

Denoting by v_n the component of $\mathbf{v}$ parallel to $\hat{\mathbf{n}}$, and by $\mathbf{v}_\tau$ the $(d - 1)$-dimensional vector in the hyper-plane perpendicular to $\hat{\mathbf{n}}$, the above equation is rewritten as

$$W_{tr}(\mathbf{V}'|\mathbf{V}) = \chi \int dv_n d\mathbf{v}_\tau \Theta[(\mathbf{V} - \mathbf{v}) \cdot \hat{\omega}] \frac{|(\mathbf{V} - \mathbf{v}) \cdot \hat{\omega}|}{\Delta v^{d-1}} p(\mathbf{v})\delta\left(\Delta V + k(\varepsilon, r)|(\mathbf{V} - \mathbf{v}) \cdot \hat{\mathbf{n}}|\right). \tag{4.11}$$

Finally, integrating over dv_n, one gets the following formula:

$$W_{tr}(\mathbf{V}'|\mathbf{V}) = \frac{1}{k(\varepsilon, r)^2}\chi|\Delta V|^{2-d}\int d\mathbf{v}_\tau \, p[\mathbf{u}(\mathbf{V}, \mathbf{V}', \mathbf{v}_\tau)], \tag{4.12}$$

where the integral in the above expression is $(d - 1)$-dimensional. The vectorial function $\mathbf{u}()$ is defined as

$$\mathbf{u}(\mathbf{V}, \mathbf{V}', \mathbf{v}_\tau) = v_n(\mathbf{V}, \mathbf{V}')\hat{\mathbf{n}}(\mathbf{V}, \mathbf{V}') + \mathbf{v}_\tau. \tag{4.13}$$

If $P(\mathbf{v}) = \frac{1}{(2\pi T)^{d/2}}\exp\left(-\frac{v^2}{2T}\right)$, the transition rate $W_{tr}(\mathbf{v}, \mathbf{v}')$ immediately follows:

$$W_{tr}(\mathbf{V}'|\mathbf{V}) = \left(\frac{1}{k(\varepsilon, r)}\right)^2 \chi \, |\Delta V|^{2-d} \frac{1}{\sqrt{2\pi T}} e^{-\frac{v_n^2}{2T}}. \tag{4.14}$$

In the following I specialize to the two dimensional case, where the above equation simplifies to

$$
W_{tr}(\mathbf{V}'|\mathbf{V}) = \chi \frac{1}{\sqrt{2\pi T/mk(\varepsilon, r)^2}}
$$
$$
\times \exp\left\{-m\left[V_n' - V_n + k(\varepsilon, r)V_n\right]^2 / (2Tk(\varepsilon, r)^2)\right\}. \tag{4.15}
$$

As discussed in details in Chap. 5, with the assumption of well-separated characteristic time scales, the dynamics of the tracer alone is Markovian, and the transition rates (which do not take into account the external driving) satisfy detailed balance with respect to a Gaussian invariant probability $P(\mathbf{V})$ [22, 27]. A simple explanation for such a counter-intuitive effect (since we are out of equilibrium) is given in Chap. 5.

When Sonine corrections are considered, one has $p(\mathbf{v}) = \frac{1}{(2\pi T/m)^{d/2}} \exp\left(-\frac{mv^2}{2T}\right)$ $(1 + a_2 S_2^d(mv^2/2T))$ with $S_2^d(x) = \frac{1}{2}x^2 - \frac{d+2}{2}x + \frac{d(d+2)}{8}$. The calculation of the integral needed to have an explicit expression of the transition rate is straightforward:

$$
\int d\mathbf{v}_{2\tau}\, p(\mathbf{v}) = \frac{e^{-\frac{mv_n^2}{2T}}}{\sqrt{2\pi T/m}}\left(1 + a_2 S_2^{d=1}(mv_n^2/2T)\right), \tag{4.16}
$$

which leads to

$$
W_{tr}(\mathbf{V}'|\mathbf{V}) = \left(\frac{1}{k(\varepsilon, r)}\right)^2 \chi \, |\Delta V|^{2-d} \frac{1}{\sqrt{2\pi T/m}} e^{-\frac{mv_n^2}{2T}} \left(1 + a_2 S_2^{d=1}\left(\frac{mv_n^2}{2T}\right)\right). \tag{4.17}
$$

Detailed balance is no more satisfied in such a case [27].

4.2 The Large Mass Limit

With the assumption of separation of time-scales discussed above, the system of Eq. (4.1) is decoupled. This allows us to write the following Master Equation for the tracer

$$
\frac{\partial P(\mathbf{V}, t)}{\partial t} = L_{gas}[P(\mathbf{V}, t)] + L_{bath}[P(\mathbf{V}, t)], \tag{4.18}
$$

and the Markovian linear operator L_{gas} can be expanded as

$$
L_{gas}[P(\mathbf{V}, t)] = \sum_{n=1}^{\infty} \frac{(-1)^n \partial^n}{\partial V_{j_1} \dots \partial V_{j_n}} \left[D_{j_1 \dots j_n}^{(n)}(\mathbf{V}) P(\mathbf{V}, t)\right], \tag{4.19}
$$

(the sum over repeated indices is meant) with

$$D^{(n)}_{j_1 \dots j_n}(\mathbf{V}) = \frac{1}{n!} \int d\mathbf{V}'(V'_{j_1} - V_{j_1}) \dots (V'_{j_n} - V_{j_n}) W_{tr}(\mathbf{V}'|\mathbf{V}), \qquad (4.20)$$

and

$$L_{bath}[P(\mathbf{V}, t)] = \mathscr{B}_{tr} P(\mathbf{V}, t). \qquad (4.21)$$

In the limit of large mass M, i.e. small ε, one expects that the interaction between the granular gas and the tracer can be described by means of an effective Langevin equation. In this case, I keep only the first two terms of the expansion [19, 31, 33, 36]

$$L_{gas}[P(\mathbf{V}, t)] = -\frac{\partial}{\partial V_i}[D^{(1)}_i(\mathbf{V}) P(\mathbf{V}, t)] + \frac{\partial^2}{\partial V_i \partial V_j}[D^{(2)}_{ij}(\mathbf{V}) P(\mathbf{V}, t)]. \quad (4.22)$$

A justification of this truncation, in the limit of small ε, comes from observing that terms $D^{(n)}_{j_1 \dots j_n}$ are of order ε^{2n}: this can be obtained by plugging the collision rule into (4.20).

It is useful at this point to introduce the velocity-dependent collision rate and the total collision frequency

$$\tilde{\omega}(\mathbf{V}) = \int d\mathbf{V}' W_{tr}(\mathbf{V}'|\mathbf{V}), \qquad (4.23)$$

$$\omega_c = \int d\mathbf{V} \, P(\mathbf{V}) \tilde{\omega}(\mathbf{V}). \qquad (4.24)$$

The former quantity can be exactly calculated under the assumption of Gaussian $p(v)$, giving

$$\tilde{\omega}(\mathbf{V}) = \chi \sqrt{\frac{\pi}{2}} \left(\frac{T}{m}\right)^{1/2} e^{-\varepsilon^2 q^2/4}$$

$$\times \left[(\varepsilon^2 q^2 + 2) I_0 \left(\frac{\varepsilon^2 q^2}{4}\right) + \varepsilon^2 q^2 I_1 \left(\frac{\varepsilon^2 q^2}{4}\right) \right], \qquad (4.25)$$

where the rescaled variable $\mathbf{q} = \mathbf{V}/\sqrt{T/M}$ is introduced in Appendix A through Eq. (A.14) and $I_n(x)$ are the modified Bessel functions. To have an approximation of ω_c, on the other side, one has to make a position about $P(\mathbf{V})$. For the sake of obtaining a first result, let us take it to be a Gaussian with variance T_{tr}/M. With this assumption, the collision rate turns out to be

$$\omega_c = \chi \sqrt{2\pi} \sqrt{T/m + T_{tr}/M} = \chi \sqrt{2\pi} \left(\frac{T}{m}\right)^{1/2} \sqrt{1 + \frac{T_{tr}}{T} \varepsilon^2} = \omega_0 K(\varepsilon), \quad (4.26)$$

where $\omega_0 = \chi \sqrt{2\pi} \left(\frac{T}{m}\right)^{1/2}$ and $K(\varepsilon) = \sqrt{1 + \frac{T_{tr}}{T}\varepsilon^2}$.

One is then able to compute the terms $D_i^{(1)}$ and $D_{ij}^{(2)}$ appearing in L_{gas}. The result and the details of the computation of these coefficients as functions of ε are given in Appendix A. Here, in order to be consistent with the approximation in (4.22), from Eq. (A.15) I report only terms up to $\mathcal{O}(\varepsilon^4)$

$$D_x^{(1)} = -\omega_0(1+r)\varepsilon^2 V_x + \mathcal{O}(\varepsilon^5) \tag{4.27}$$

$$D_y^{(1)} = -\omega_0(1+r)\varepsilon^2 V_y + \mathcal{O}(\varepsilon^5) \tag{4.28}$$

$$D_{xx}^{(2)} = D_{yy}^{(2)} = \chi\sqrt{\pi/2}\left(\frac{T}{m}\right)^{3/2}(1+r)^2\varepsilon^4 + \mathcal{O}(\varepsilon^5)$$

$$= \frac{\omega_0}{2}\frac{T}{m}(1+r)^2\varepsilon^4 + \mathcal{O}(\varepsilon^5) \tag{4.29}$$

$$D_{xy}^{(2)} = \mathcal{O}(\varepsilon^6). \tag{4.30}$$

The linear dependence of $D_\beta^{(1)}$ upon V_β (for each Cartesian component β) allows to define an effective granular linear drag with coefficient

$$\eta_g = \omega_0(1+r)\varepsilon^2. \tag{4.31}$$

In the elastic limit $r \to 1$, one retrieves the classical results: $\eta_g \to 2\omega_0\varepsilon^2$ and $D_{xx}^{(2)} = D_{yy}^{(2)} \to 2\omega_0\varepsilon^2\frac{T}{M}$. In this limit the Fluctuation-Dissipation relation of the second kind is satisfied [20, 21, 26], i.e. the ratio between the noise amplitude and η_g, associated to the same source (collision with gas particles), is exactly T/M. When the collisions are inelastic, $r < 1$, one sees two main effects: 1) the time scale associated to the drag $\tau_g = 1/\eta_g$ is modified by a factor $\frac{1+r}{2}$, i.e. it is weakly influenced by inelasticity; 2) the Fluctuation-Dissipation relation of the second kind is *violated* by the same factor $\frac{1+r}{2}$. This is only a partial conclusion, which has to be re-considered in the context of the full dynamics, including the external bath: this is discussed below, in Chap. 5 [16].

4.2.1 Langevin Equation for the Tracer

Putting together the results in Eqs. (4.27–4.30) with Eqs. (4.18–4.22), a Langevin equation for the tracer can be written:

$$M\dot{\mathbf{V}} = -\Gamma\mathbf{V} + \mathcal{E}, \tag{4.32}$$

where $\Gamma = \gamma_b + \gamma_g$ and $\mathcal{E} = \xi_b + \xi_g$, with

$$\gamma_g = M\eta_g = M\omega_0(1+r)\varepsilon^2 = \omega_0(1+r)m \tag{4.33}$$

$$\langle \mathscr{E}_i(t)\mathscr{E}_j(t')\rangle = 2\left[\gamma_b T_b + \gamma_g \left(\frac{1+r}{2}T\right)\right]\delta_{ij}\delta(t-t'), \qquad (4.34)$$

concluding that the stationary velocity distribution of the intruder is Gaussian with temperature

$$T_{tr} = \frac{\gamma_b T_b + \gamma_g \left(\frac{1+r}{2}T\right)}{\gamma_b + \gamma_g}. \qquad (4.35)$$

Equation (4.32) is consistent with the Gaussian ansatz used in computing ω_0.

I resume here the main consequences of Eq. (4.32), specializing for simplicity to the one-dimensional case:

$$M\dot{V} = -\Gamma V + \mathscr{E} \qquad (4.36)$$

with $\langle \mathscr{E}\rangle = 0$ and $\langle \mathscr{E}(t)\mathscr{E}(t')\rangle = 2T_{tr}\Gamma\delta(t-t')$.

The solution of this stochastic equation is

$$V(t) = e^{-t\Gamma/M}\left[V(0) + \int_0^t ds\,e^{s\Gamma/M}\mathscr{E}(s)\right] \qquad (4.37)$$

which implies, in the stationary state, that

$$C(t) = \langle V(t)V(0)\rangle = \langle V^2\rangle e^{-t\Gamma/M}. \qquad (4.38)$$

The position of the intruder follows the diffusion equation which implies

$$\langle (X(t)-X(0))^2\rangle = \left\langle \int_0^t ds\int_0^t ds'\,V(s)V(s')\right\rangle \to 2Dt \quad (t\to\infty) \qquad (4.39)$$

with

$$D = \int_0^\infty dt\,C(t). \qquad (4.40)$$

For the granular intruder it is immediately obtained

$$D = \frac{T_{tr}}{\Gamma} = \frac{\gamma_b T_b + \gamma_g \left(\frac{1+r}{2}T\right)}{(\gamma_b + \gamma_g)^2}. \qquad (4.41)$$

Solving numerically the equation for the granular temperature and substituting the result into the above equation, one can study D as a function of the restitution coefficient r [33]. When all other parameters are kept constant and r is reduced from 1, the behavior of D is non-monotonic, it decreases, has a minimum and then increases for lower values of r. Anyway, this minimum is expected for quite low values of r or high values of the packing fraction ϕ, where the approximations involved in this theory are not good. For this reason, at the values of parameters chosen to have a good comparison with simulations, this non-monotonic behavior is not observed.

It should be also noticed that, in the Homogeneous Cooling State, the self-diffusion coefficient at a given granular temperature increases as r is reduced from 1, i.e. it has an opposite behavior with respect to the present case [1, 2]. Other studies on different models of driven granular gases have found expressions very close to Eq. (4.33), which is not surprising considering the universality of the main ingredient for this quantity, i.e. the collision integral [4, 25].

4.3 Non-Markovian Tracer's Diffusion

As the packing fraction is increased, the Enskog approximation fails in predicting dynamical properties [16, 29]. In particular, the velocity autocorrelation function (VACF) $C(t) = \langle V(t)V(0) \rangle$ shows an exponential decay modulated by oscillating functions [12, 33]. The Enskog approximation is unable to explain the observed functional forms, because it only modifies by a constant factor the collision frequency [3, 33]: a model with more than one characteristic time is needed. Additional time-scales appear if memory effects are considered, therefore a non-Markovian model is needed. However, as it is often the case when memory decays in a *finite* time, a non-Markovian model can be mapped onto a Markovian one by increasing the number of degrees of freedom. Very slow time-decay of memory kernels apparently rule out such a Markovian embedding: nevertheless, even non-integrable power-law decays can be represented as sum of a few exponential decays [15, 28] up to a time-scale which is sufficiently long for all practical purposes.

A first approximation beyond the Markovian limit discussed above, is given by coupling the intruder's velocity to an auxiliary field:

$$M\dot{V} = -\Gamma_E(V - U) + \sqrt{2\Gamma_E T}\,\mathscr{E}_V \qquad (4.42)$$
$$M'\dot{U} = -\Gamma' U - \Gamma_E V + \sqrt{2\Gamma' T_b}\,\mathscr{E}_U,$$

where $\mathscr{E}_V$ and $\mathscr{E}_U$ are white noises of unitary variance. Two new parameters appear: the mass of the local field M' and its drag coefficient Γ'. The dilute limit here is obtained for $\Gamma' \sim M' \to \infty$. In such a limit indeed $U \to 0$ and the equation for V comes back in the form discussed above [33], see Eq. (4.32).

In Eq. (4.42), the dynamics of the tracer is remarkably simple: indeed V follows a Langevin equation in a *Lagrangian frame* with respect to a field U, which can be interpreted as the *local average velocity field* of the gas particles colliding with the tracer. A first justification of this model comes from realizing [28, 38] that it is equivalent to a Generalized Langevin Equation with exponential memory, which is consistent with a typical approximation done for Brownian Motion when, at high densities, the coupling of the intruder with fluid hydrodynamic modes, decaying exponentially in time (see [39], Sects. 8.6 and 9.1), must be taken into account. Here such a coupling, which in principle involves a continuum of modes, is simplified to be dominated by a single mode. This is sufficient to introduce a new timescale which explains the oscillations in the VACF.

The full coupling would reproduce finer features which become relevant at larger densities or larger times, such as long-time power-law tails. The fact that the "temperature" of the local velocity field U is equal to the bath temperature T_b comes as a consequence of the conservation of momentum in collisions, implying that the average velocity of a group of particles is not changed by collisions among themselves and is only affected by the external bath and a (small) number of collisions with outside particles. This scenario is fully consistent with the study of hydrodynamic fluctuations for the velocity field of the same fluid model [17, 18].

A stronger justification comes, however, from its effectivness in reproducing the numerical results, as detailed in [33]. From the simulations it is seen that the relaxation time of the local field $\tau_U = M'/\Gamma'$, rescaled by the mean collision time, increases with the packing fraction and with the inelasticity, as expected. At high densities it appears that $\Gamma' \sim 1/\phi$, and $T_{tr} \sim T$, likely due to stronger correlations among particles. At large ϕ, moreover, T_{tr} is larger than the value predicted with Molecular Chaos, Eq. (4.35), consistently with a *smaller* dissipation for correlated collisions.

Equation (4.42) is not only able to reproduce the non-monotonic VACF seen in numerical experiments, but also explains the violation of the Einstein relation, which is one of the non-equilibrium effects discussed in Chap. 5. The presence of an additional degree of freedom, $U(t)$, is sufficient to break detailed balance. Such a possibility was absent in the Langevin model, valid in the dilute limit, Eq. (4.32).

4.4 The Granular Brownian Ratchet

The topic of the intruder considered in Sect. 4.1 has an interesting extension, when an *asymmetric* shape is considered. An example is shown in Fig. 4.1.

The model depicted in the Figure and discussed in this Section [6, 7], consists of a triangular particle of mass M, shaped as an isosceles triangle with base l and angle opposite to the base $2\theta_0$ and surrounded by a gas of N disks of diameter $\sigma = 1$ and mass $m = 1$. For analogy with other mechanisms [30], in the following I call "ratchet" the triangular particle. Note that the ratchet can only slide, without rotating, along the direction x, perpendicular to its base and the whole system is enclosed in a squared box of side L with periodic boundary conditions. The $N + 1$ particles (for the moment all denoted by velocity $\mathbf{v}$) undergo binary instantaneous collisions described by the rule:

$$\mathbf{v}_i = \mathbf{v}'_i - (1 + r_{ij})c_{ij}[(\mathbf{v}'_i - \mathbf{v}'_j) \cdot \hat{\mathbf{n}}]\hat{\mathbf{n}}, \qquad (4.43)$$

Fig. 4.1 Sketch of the 2D model. The *triangle* is constrained to move only in the $\hat{x}$ (*left/right*) direction, while its orientation is fixed, i.e. it cannot rotate. Gas particles collide against it and occasionally receive energy from an external bath

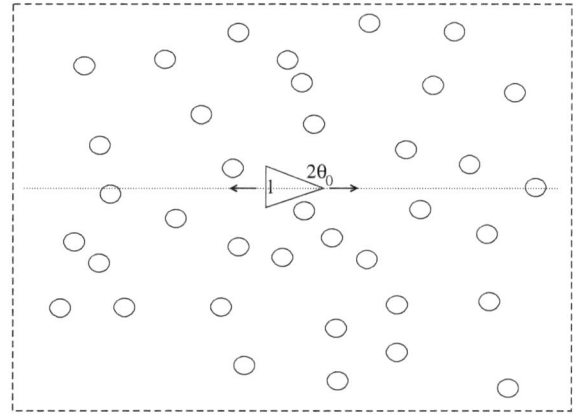

where $\mathbf{v}$ and $\mathbf{v}'$ are the post-collisional and pre-collisional velocities, respectively. The quantity $r_{ij} \leq 1$ are the coefficients of restitution for that particular collision, taking value r_d if both objects are disks or value r_r if the ratchet is involved, $\hat{n}$ is the outward-pointing unit vector normal, in the contact point, to the surface of particle i, and c_{ij} is a coefficient which takes, in the different collisions, the values

$$c_{ij} = \begin{cases} 1/2 & \text{if objects are both disks} \\ 1/(1 + \varepsilon^2 \hat{n}_x^2) & \text{if } j \text{ is the triangle} \\ \varepsilon^2/(1 + \varepsilon^2 \hat{n}_x^2) & \text{if } i \text{ is the triangle} \end{cases} \qquad (4.44)$$

where $\varepsilon^2 = m/M$. Because of the constraint, the vertical velocity of the ratchet is always 0. The collision rule (4.43) conserves the total momentum if i and j are disks, and conserves the x-component of the momentum only, when the triangle is involved. If $r_{ij} = 1$ the total kinetic energy is also conserved. Three possible cases may be considered: (i) a pure elastic gas where $r_d = r_r = 1$, (ii) a mixed gas where $r_d = 1$ and $r_r < 1$, (iii) a pure inelastic gas where $r_d < 1$ and $r_r < 1$. In both cases (ii) and (iii) an external driving mechanism is needed to attain a stationary state and avoid indefinite cooling of the system. To simplify the discussion, the driving mechanism is again assumed to be analogous to a thermostat as in Sect. 2.3.5, and—in particular— it is coupled only to the gas particles. The ratchet reaches a statistically steady state because of the coupling with the gas.

4.4.1 Continuous Limit

In the Markovian limit discussed in Sect. 4.1, it is reasonable to study the ratchet dynamics by means of the first of Eq. (4.1). Having detached the intruder from the heat bath, I put $\mathscr{B}_{tr} \equiv 0$. The intruder's transition rate now includes the details of the collision which depend on the point (angle θ) along the intruder's surface, i.e.

$$W_{tr}(V|V') = n \int_0^{2\pi} d\theta\, SF(\theta) \int_{-\infty}^{\infty} dv'_x \int_{-\infty}^{\infty} dv'_y\, p(v'_x, v'_y)$$
$$\times (\mathbf{V}' - \mathbf{v}') \cdot \hat{\mathbf{n}} \Theta[(\mathbf{V}' - \mathbf{v}') \cdot \hat{\mathbf{n}}] \cdot \delta[V - V_{post}(V', \mathbf{v}', r_r, \varepsilon)] \quad (4.45)$$

with $p(\mathbf{v})$ the gas particle distribution, V_{post} the post-collisional ratchet velocity [see Eq. (4.43)], Θ the Heaviside step function, S the perimeter length, $\hat{\mathbf{n}} =$
$(\sin\theta, -\cos\theta)$ and for the triangle

$$SF(\theta) = \frac{l}{2\sin\theta_0}\{2\sin\theta_0\delta(\theta - 3\pi/2) + \delta(\theta - \theta_0) + \delta[\theta - (\pi - \theta_0)]\}. \quad (4.46)$$

Following numerical evidence, I approximate the velocity pdf of the gas, $p(\mathbf{v})$, by
a Maxwellian with zero mean and variance T. It is straightforward to verify that
detailed balance, in the form

$$P(V)W_{tr}(V'|V) = P(-V')W_{tr}(-V| - V'), \quad (4.47)$$

holds only if $r_r = 1$.

As numerical results suggest, the ME describes a driven-diffusive process. In order
to gain a deeper insight, it is convenient to approximate the ME by a Fokker-Planck
equation (FPE), from which I can extract the analytical expression of the drift and
diffusion terms. This is achieved by expanding the transition rates as in Sect. 4.2.
By retaining only the first two terms one obtains the sought FPE, which can be
still simplified by expanding these terms in the small parameter ε. The resulting
expressions suggest a simple physical picture, which can be illustrated with the help
of the Langevin equation associated with the FPE:

$$\dot{V}(t) = -\gamma V(t) + \frac{F}{M} + \mathscr{E}(t) \quad (4.48)$$

with noise

$$\langle \mathscr{E}(t)\mathscr{E}(t')\rangle = \frac{2\gamma T_r}{M}\delta(t - t') \quad \langle \mathscr{E}(t)\rangle = 0 \quad (4.49)$$

The quantities γ and F are effective parameters related to the original parameters by

$$\gamma = 4\eta nl\varepsilon\sqrt{\frac{T}{2\pi M}}(1 + \sin\theta_0) \quad (4.50)$$

$$\frac{F}{M} = -nl\frac{T}{M}\varepsilon^2(1 - \sin^2\theta_0)\eta(1 - \eta) \quad (4.51)$$

$$1 - \eta = 1 - \frac{T_r}{T} = \frac{1 - r_r}{2} \tag{4.52}$$

Hence, for $r_r < 1$ the ratchet drifts with an average negative velocity

$$\langle V(t) \rangle = \frac{F}{M\gamma} = -\frac{1 - r_r}{8} \sqrt{\frac{2\pi T}{M}} \varepsilon (1 - \sin \theta_0). \tag{4.53}$$

Indeed, the net velocity vanishes linearly with $\varepsilon \to 0$ and is very tiny for massive ratchets. It is of interest to observe that in virtue of Eq. (4.52) the net driving force is proportional to the temperature difference $T - T_r$, so that the tracer and the gas temperatures play roles analogous to the two reservoir temperatures of the Brownian ratchet model considered by [37]. In principle it is possible that for a purely inelastic system (case iii), for some choice of inelasticity and masses, the difference $T - T_r$ can change sign, implying a change of sign of the average ratchet velocity.

From Eqs. (4.48)–(4.52) it is also possible to estimate the signal to noise ratio:

$$\sqrt{\frac{\langle V(t) \rangle^2}{\langle V^2(t) \rangle - \langle V(t) \rangle^2}} \simeq \sqrt{2\pi} \frac{1 - r_r}{8} \varepsilon (1 - \sin \theta_0). \tag{4.54}$$

The measure of $\langle V \rangle$ can be blurred by thermal noise in the limit of large M/m, a fact that can be avoided with a large number of independent trajectories.

4.4.2 Other Methods, Models and Experiments

Collisional granular ratchets, or "motors", have been studied under different kinds of assumptions and with several models. The motor effect is present (it can even be much stronger) also when the mass of the intruder is comparable or even smaller than the mass of the gas particles. In such a situation the computation for $\varepsilon \ll 1$ does not reproduce the numerical results. Typically skewed non-Gaussian distributions $P(\mathbf{V})$ appear and other approximations must be used [8].

The signal-to-noise ratio for the motor effect is considerably improved in the case of a symmetric intruder with opposite *faces* made of different materials, i.e. with different restitution coefficient [8]. In that case it is found that the $\langle V \rangle$ is of order $\sqrt{T_{tr}/M}$ instead of $\varepsilon \sqrt{T_{tr}/M}$.

Experiments have also been performed in order to reproduce the predicted effect [10, 13, 24]. In a real experiment, however, among the many effects not included in the above theory, one plays a particular role: Coulomb friction. To suspend the intruder in the granular fluid it is necessary to put it in contact with some bearings. In most of the cases the intruder will suffer from an additional *dissipative* force: $F_{coul} \propto \text{sign}(V)$ where sign() is the sign function. The theoretical study of the effect of Coulomb friction on the model (4.1), even in the Markovian limit, is in its infancy, but it has already revealed interesting surprises [9, 14, 32, 34, 35].

References

1. Brey, J.J., Dufty, J.W., Santos, A.: Kinetic models for granular flow. J. Stat. Phys. **97**, 281 (1999)
2. Brey, J.J., Ruiz-Montero, M.J., Garcia-Rojo, R., Dufty, J.W.: Brownian motion in a granular gas. Phys. Rev. E **60**, 7174 (1999)
3. Brilliantov, N.V., Pöschel, T.: Kinetic Theory of Granular Gases. Oxford University Press, Oxford (2004)
4. Bunin, G., Shokef, Y., Levine, D.: Frequency-dependent fluctuation-dissipation relations in granular gases. Phys. Rev. E **77**, 051301 (2008)
5. Chapman, S., Cowling, T.G.: The Mathematical Theory of Nonuniform Gases. Cambridge University Press, Cambridge (1960)
6. Cleuren, B., Van den Broeck, C.: Granular brownian motor. Europhys. Lett. **77**, 50003 (2007)
7. Costantini, G., Puglisi, A., Marini Bettolo Marconi, U.: A granular brownian ratchet model. Phys. Rev. E **75**, 061124 (2007)
8. Costantini, G., Marconi, U.M.B., Puglisi, A.: Noise rectification and fluctuations of an asymmetric inelastic piston. Europhys. Lett. **82**, 50008 (2008)
9. de Gennes, P.G.: Brownian motion with dry friction. J. Stat. Phys. **119**, 953 (2005)
10. Eshuis, P., van der Weele, K., Lohse, D., van der Meer, D.: Experimental realization of a rotational ratchet in a granular gas. Phys. Rev. Lett. **104**, 248001 (2010)
11. Feitosa, K., Menon, N.: Breakdown of energy equipartition in a 2D binary vibrated granular gas. Phys. Rev. Lett. **88**, 198301 (2002)
12. Fiege, A., Aspelmeier, T., Zippelius, A.: Long-time tails and cage effect in driven granular fluids. Phys. Rev. Lett. **102**, 098001 (2009)
13. Gnoli, A., Petri, A., Dalton, F., Gradenigo, G., Pontuale, G., Sarracino, A., Puglisi, A.: Brownian ratchet in a thermal bath driven by Coulomb friction. Phys. Rev. Lett. **110**, 120601 (2013)
14. Gnoli, A., Puglisi, A., Touchette, H.: Granular brownian motion with dry friction. Europhys. Lett. **102**, 14002 (2013)
15. Goychuk, I.: Viscoelastic subdiffusion: generalized Langevin Equation approach. Adv. Chem. Phys. **150**, 187 (2012)
16. Gradenigo, G., Puglisi, A., Sarracino, A., Villamaina, D.: Irreversible dynamics of a massive intruder in dense granular fluids. Europhys. Lett. **92**, 34001 (2010)
17. Gradenigo, G., Sarracino, A., Villamaina, D., Puglisi, A.: Fluctuating hydrodynamics and correlation lengths in a driven granular fluid. J. Stat. Mech. **8**, P08017 (2011)
18. Gradenigo, G., Sarracino, A., Villamaina, D., Puglisi, A.: Non-equilibrium length in granular fluids: from experiment to fluctuating hydrodynamics. Europhys. Lett. **96**, 14004 (2011)
19. van Kampen, N.G.: A power series expansion of the master equation. Can. J. Phys. **39**, 551 (1961)
20. Kubo, R., Toda, M., Hashitsume, N.: Statistical Physics II. Nonequilibrium Stastical Mechanics. Springer, Berlin (1991)
21. Marconi, U.M.B., Puglisi, A., Rondoni, L., Vulpiani, A.: Fluctuation-dissipation: response theory in statistical physics. Phys. Rep. **461**, 111 (2008)
22. Martin, P.A., Piasecki, J.: Thermalization of a particle by dissipative collisions. Europhys. Lett. **46**, 613 (1999)
23. Pagnani, R., Marconi, U.M.B., Puglisi, A.: Driven low density granular mixtures. Phys. Rev. E **66**, 051304 (2002)
24. Gnoli, A., Sarracino, A., Petri, A., Puglisi, A.: Nonequilibrium fluctuations in frictional granular motor: experiments and kinetic theory. Phys. Rev. E. **87**, 052209 (2013)
25. Puglisi, A., Baldassarri, A., Vulpiani, A.: Violations of the Einstein relation in granular fluids: the role of correlations. J. Stat. Mech. P08016 (2007)
26. Puglisi, A., Baldassarri, A., Loreto, V.: Fluctuation-dissipation relations in driven granular gases. Phys. Rev. E **66**, 061305 (2002)
27. Puglisi, A., Visco, P., Trizac, E., van Wijland, F.: Dynamics of a tracer granular particle as a nonequilibrium Markov process. Phys. Rev. E **73**, 021301 (2006)

28. Puglisi, A., Villamaina, D.: Irreversible effects of memory. Europhys. Lett. **88**, 30004 (2009)
29. Puglisi, A., Sarracino, A., Gradenigo, G., Villamaina, D.: Dynamics of a massive intruder in a homogeneously driven granular fluid. Gran. Matt. **14**, 235 (2012)
30. Reimann, P.: Brownian motors: noisy transport far from equilibrium. Phys. Rep. **361**, 57 (2002)
31. Risken, H.: The Fokker-Planck equation: methods of solution and applications. Springer, Berlin (1989)
32. Sarracino, A., Manacorda, A., Puglisi, A.: Coulomb friction driving brownian motors. Comm. Theor. Phys. xxx:xxx (2014) (in press)
33. Sarracino, A, Villamaina, D., Costantini, G., Puglisi, A.: Granular brownian motion. J. Stat. Mech. P04013 (2010)
34. Sarracino, A.: Time asymmetry of the Kramers equation with nonlinear friction: fluctuation-dissipation relation and ratchet effect. Phys. Rev. E **88**, 052124 (2013)
35. Sarracino, A., Gnoli, A., Puglisi, A.: Ratchet effect driven by Coulomb friction: the asymmetric Rayleigh piston. Phys. Rev. E **87**, 040101(R) (2013)
36. Spohn, H.: Kinetic equations from Hamiltonian dynamics: Markovian limits. Rev. Mod. Phys. **52**, 569 (1980)
37. Van den Broeck, C., Kawai, R.: Microscopic analysis of a thermal brownian motor. Phys. Rev. Lett. **93**, 090601 (2004)
38. Villamaina, D., Baldassarri, A., Puglisi, A., Vulpiani, A.: Fluctuation dissipation relation: how does one compare correlation functions and responses? J. Stat. Mech. P07024 (2009)
39. Zwangzig, R.: Nonequilibrium statistical mechanics. Oxford Univ, Press (2001)

Chapter 5
The Arrow of Time: Past and Future of Grains

Abstract In a granular fluid the balance between energy fluxes entering and leaving the system establishes a non-equilibrium stationary state. Therefore time-reversal symmetry is broken and this affects the statistical features of many observables. A few examples are discussed here, pertaining to the two main paradigms studied in previous chapters: tracer's dynamics and hydrodynamics. In both cases the choice of a reduced number of degrees of freedom appears as a contraction of information. Sometimes this reduction affects non-equilibrium properties.

5.1 Equilibrium from a Dynamical Perspective

The condition of thermodynamic equilibrium can be stated in a very general way, from the point of view of dynamics. Let us assume that we have an evolution equation which generates trajectories between time 0 and time t, and let us denote each trajectory as

$$\Omega_0^t \equiv \{\mathbf{r}(s), \mathbf{v}(s)\}_{s=0}^t \tag{5.1}$$

starting from an initial condition $\mathbf{r}(0), \mathbf{v}(0)$. The evolution can be stochastic (as in Eq. (4.32)) or deterministic with random initial conditions. In both cases we have an ensemble of possible trajectories. Let us assume that we are somehow able to determine the probability "weight" (or density, if in a continuous space) of each Ω_0^t: $P(\Omega_0^t)$. We stress that this is meant to be the *absolute* probability, i.e. it is not conditioned by initial conditions.

Finally, let us define the time inversion operator $\mathcal{T}$

$$\mathcal{T}\Omega_0^t = \{\mathbf{r}(t-s), -\mathbf{v}(t-s)\}_{s=0}^t. \tag{5.2}$$

Obviously $\mathcal{T}^2 = \mathcal{I}$ (the identity operator).

The condition of thermodynamic equilibrium is equivalent to dynamical time reversibility [29]. It simply states that

$$P(\Omega_0^t) = P(\mathcal{T}\Omega_0^t). \tag{5.3}$$

© The Author(s) 2015   97
A. Puglisi, *Transport and Fluctuations in Granular Fluids*,
SpringerBriefs in Physics, DOI 10.1007/978-3-319-10286-3_5

5.1.1 The Case of Markov Processes

For a continuous time Markov process $\sigma(t)$, a trajectory is described by the sequence of visited states $(\sigma_0, \sigma_1, \sigma_2, ..., \sigma_n)$ and the time of permanence in each state $(t_0, t_1, t_2, ..., t_n$ with $\sum t_i = t)$ and its probability is

$$P(\Omega_0^T) = p(\sigma_0, 0) p_{perm}(\sigma_0, t_0) W(\sigma_0 \to \sigma_1) p_{perm}(\sigma_1, t_1)...W(\sigma_{n-1} \to \sigma_n)$$
$$\times p_{perm}(\sigma_n, t - t_{n-1}), \tag{5.4}$$

where $p(\sigma, t)$ is the probability of finding the process in state σ at time t, $p_{perm}(\sigma, t)$ is the probability of staying for a time t in state σ, and $W(\sigma \to \sigma')$ is the conditional probability of changing state from σ to σ'.

In this case, condition (5.3) is satisfied if and only if

1. the system is in the stationary state, i.e. the time is very large and any memory of the initial condition is lost; at large times one has $p(\sigma, t) \to \mu(\sigma)$ the so-called invariant probability;
2. for any couple of states σ and σ', the transition rates and the invariant measure must satisfy the following condition

$$\mu(\sigma) W(\sigma \to \sigma') = \mu(\mathscr{T}\sigma') W(\mathscr{T}\sigma' \to \mathscr{T}\sigma), \tag{5.5}$$

called "detailed balance condition".

5.1.2 Entropy Production

Whenever condition (5.3) is not satisfied, one may distinguish between the forward and the backward time direction by appropriate measurements. Spatially extended systems reveal their non-equilibrium properties through the appearance of spatially directed currents. For instance, a substance coupled to two thermostats at different temperatures $T_1 > T_2$, is crossed by a heat current flowing from temperature T_1 to T_2. Anyway, the definition of equilibrium based on the probability of trajectories, allows to construct a more general and abstract "current" which universally reveals the presence of a time-arrow:

$$J_t = \lim_{t \to \infty} \frac{1}{t} \langle \mathscr{W}_t \rangle \tag{5.6}$$

with

$$\mathscr{W}_t = \log \frac{P(\Omega_0^t)}{P(\mathscr{T}\Omega_0^t)}. \tag{5.7}$$

The current defined in (5.6) is called "entropy production rate" and the *stochastic variable* in (5.7) is said "fluctuating entropy production". This latter quantity has been extensively analyzed in [15].

It is immediate to see that—in the stationary state—one has

$$\frac{\text{prob}(\mathscr{W}_t = x)}{\text{prob}(\mathscr{W}_t = -x)} = e^x \tag{5.8}$$

which is called "fluctuation relation" (or sometimes "transient fluctuation relation"). This is a simpler version of a very general relation which is valid (with appropriate definitions of its object) also for chaotic deterministic systems [6–8].

For Markov processes one easily finds

$$\mathscr{W}_t = \log \frac{p(\sigma_0, 0)}{p(\sigma_n, t)} + \sum_{i=0}^{n-1} \log \frac{W(\sigma_i \to \sigma_{i+1})}{W(\sigma_{i+1} \to \sigma_i)} \approx \sum_{i=0}^{n-1} \log \frac{W(\sigma_i \to \sigma_{i+1})}{W(\sigma_{i+1} \to \sigma_i)} \tag{5.9}$$

where the last approximation is true for bounded systems and large times (for large times, when the system is not bounded, it may be not true, see for instance the discussion in [21], and references therein).

5.1.3 Observables Related to Entropy Production

It is clear that the entropy production defined in (5.7) is very difficult to be measured as it is: even in simulations, one needs an expression for $P(\Omega_0^t)$ which is not easily calculated for a generic process. This is simplified for Markov processes, but the problem of "experimentally" accessing (5.7) remains open, when a model (e.g. transition rates) is not available [34]. In many situations, particularly those with a well defined thermostat at temperature k_B/β, it is found that entropy production is related to the power injected by non-conservative forces acting on the system, e.g.

$$J_t \approx \beta \dot{w}_{nc}, \tag{5.10}$$

where w_{nc} is the work of non-conservative forces. Such a work is often given as product of a generalized force and an internal current generated by the force (for instance a difference of potential generating a charge current). Unfortunately this relation is not as general as one hopes: it is sufficient to realize that there are many non-equilibrium situations where a temperature is not clearly defined. Relation (5.10) is considered to be valid in all situations *near equilibrium*, where—for instance—the so-called "non-equilibrium thermodynamics" fairly describes the system [4] and entropy production has a definition in terms of thermodynamic currents and generalized thermodynamic forces.

An instructive example of calculation of the entropy production can be given for a simple process which is a slight generalization of Eq. (4.36):

$$\dot{v} = -\Gamma v + F(t) + \mathscr{E} \tag{5.11}$$

with Gaussian noise $\langle \mathscr{E} \rangle = 0$, $\langle \mathscr{E}(t)\mathscr{E}(t') \rangle = 2T\Gamma\delta(t - t')$, and where $F(t) = F_c + F_{nc}$ is a sum of a conservative force $F_c = -U'(x)$ and a non-conservative force $F_{nc}(t)$. This is also the equation that governs, at a first approximation, the process of pulling a terminal of a macromolecule anchored to a surface and surrounded by water; this system has been studied in recent experiments [16, 24].

To compute the probability of a trajectory, it is sufficient to consider discrete times $t_0 + k\tau$ with τ arbitrarily small and $k \in [0, n]$ with n being the integer part of $(t - t_0)/\tau$. Since the noise is Gaussian and delta-correlated, the sequence of variables $\eta_k = \eta(t_0 + k\tau)$ has the probability density

$$P[(\eta_n, t|...|\eta_0, 0)] \propto \exp\left(-\frac{1}{2}\sum_{k=0}^{n}\eta_k^2\tau\right) \tag{5.12}$$

which, in the limit $\tau \to 0$, becomes

$$P[(\eta_n, t|...|\eta_0, 0)] \propto \exp\left(-\frac{1}{2}\int_0^t ds\,\eta^2(s)\right). \tag{5.13}$$

Equation (5.11) tells us that $\eta(t) = (\dot{v} + \Gamma v - F)/\sqrt{2\Gamma T}$, which finally gives us

$$P[\{\eta(t)\}] \propto \exp(-L), \tag{5.14}$$

where

$$L = \frac{1}{4\Gamma T}\int_0^t ds\,(\dot{v} + \Gamma v - F)^2 = \int_0^t \frac{\dot{v}^2 + \Gamma^2 v^2 + F^2 - 2F\dot{v}}{4\Gamma T}ds$$

$$+ \frac{v^2(t) - v^2(0) + 2\{U[x(t)] - U[x(0)]\}}{4T} - \frac{\int_0^t F_{nc}(s)v(s)ds}{2T} \tag{5.15}$$

is called the thermodynamic action. To find the most probable path from $(x_0, 0)$ to (x_t, t), it is sufficient to minimize the action (5.15) while keeping fixed the endpoints. The entropy production reads:

$$\mathscr{W}_t = \log\frac{P(\Omega_0^t)}{P(\mathscr{T}\Omega_0^t)} = \frac{\Delta H}{T} + \frac{\int_0^t F_{nc}(s)v(s)ds}{T} \tag{5.16}$$

where $\Delta H = \frac{v^2(t) - v^2(0)}{2} + U[x(t)] - U[x(0)]$. Equation (5.16), for large times, allows one to identify the work done by non-conservative forces $w_{nc}(t) = F_{nc}(t)v(t)$ done by the external non-conservative force (divided by T) as the entropy produced during the time t. This is an example of the result by Kurchan [13] and by Lebowitz and Spohn [15] about the Fluctuation Relation for stochastic systems. An updated review of the huge amount of literature which focuses on stochastic thermodynamics in Langevin systems can be found in [27].

5.2 The Case of the Granular Intruder

As seen in Chap. 4, when $N \gg 1$ and in the limit of vanishing packing fraction, the gas evolution is not perturbed by the intruder, which implies that the granular intruder performs a Markov process. An analysis of the transition rates $W_{tr}(V'|V)$ related to the collision with the gas [23] shows that the detailed balance property depends on the form of the $p(\mathbf{v})$ assumed for the gas. The intruder transition rates satisfy detailed balance (with respect to a Gaussian invariant probability), if $p(\mathbf{v})$ is Gaussian, otherwise detailed balance is violated.

Anyway one has to consider the combined effect of the two baths, i.e. the collisions with the gas together with the stochastic force of the external bath, as it is stated in the first of Eqs. (4.1). It is not difficult to realize that, even if $p(\mathbf{v})$ is Gaussian, this total rate cannot satisfy detailed balance. The conclusion is that, in general, the granular intruder cannot be modelled as an equilibrium process. Given in different words, one has always the possibility—measuring suitable observable—to discriminate between the correct time direction and its inverse.

5.2.1 The Paradox of the Large Mass Limit

In Chap. 4 we have seen that the granular intruder, in the limit of large mass and when the surrounding gas is dilute, follows a Langevin equation with white noise and linear drag. Only the formula for the parameters of the equation (drag coefficient Γ and noise amplitude $\langle \mathscr{E}^2 \rangle$) show the joint effect of two different baths, which is peculiar of granular systems. Anyway the linear Langevin equation is a standard example of stochastic motion. It was proposed by Paul Langevin [14]—roughly a century ago–to describe the so-called Brownian motion, i.e. the erratic trajectory of a pollen grain suspended in water. In that case the "bath" is unique (just water) and the system is at equilibrium, i.e. is invariant under the operation of time-reversal. Thanks to such a symmetry, the work of Langevin (and of Einstein before him [5]) to compute the coefficients of the equation was significantly simplified: there is no need to derive them from microscopic kinetic equations, as we have done in Sect. 4.2.

Nevertheless, apart from the difference in the derivation, the equations are the same. If one looks at the motion of the granular intruder any measurement would give

the same results as for the pollen grain: in other words, it is not possible to realize that
non-equilibrium processes occur (e.g. the inelastic collisions), if the intruder position
and velocity are the only available observables. Equivalently, Eq. (5.16) shows that
for the intruder in the large mass limit, the entropy production rate is zero.

Could we expect this result? The intruder is coupled to *two* different baths, one is
the original (external) thermostat, the second is the "gas" surrounding the intruder,
which acts as a bath in the large mass limit. What about energy fluxes in this system?

The energy injection rates of the two termostats are [33]

$$Q_b = \langle \mathbf{V}(t) \cdot (\boldsymbol{\xi}_b - \gamma_b \mathbf{V}) \rangle = 2\frac{\gamma_b}{M}(T_b - T_{tr}) \tag{5.17}$$

$$Q_g = \langle \mathbf{V}(t) \cdot (\boldsymbol{\xi}_g - \gamma_g \mathbf{V}) \rangle = 2\frac{\gamma_g}{M}(T' - T_{tr}), \tag{5.18}$$

where $T' = \frac{1+r}{2}$ is the granular temperature "probed by the tracer". It is easy to
see that the balance of fluxes $Q_b = -Q_g$ is equivalent to formula (4.35) for T_{tr}.
This balance implies that, if $T_{tr} < T_b$, then $T_{tr} > T'$. When $r < 1$, the two fluxes
are different from zero, i.e. energy is flowing from the external driving, through the
tracer, into the granular bath.

Apparently, this contradicts the "equilibrium" nature of the Langevin equation (4.32). Actually this is not a paradox but only a consequence of Molecular
Chaos and the separation of time-scales which allows us to write Eq. (4.18) without
memory terms [19, 25]. The absence of memory terms implies that both ξ_b and ξ_g are
white noises and makes them undistinguishable: an observer which can only measure
$\mathbf{V}(t)$ cannot obtain separate measures of Q_b and Q_g, but only a measure of the total
energy flow $Q = M\langle \mathbf{V} \cdot \dot{\mathbf{V}} \rangle = 0$ which hides out the presence of energy currents.

A more detailed analysis, e.g. by relaxing some of the assumptions (large mass,
infinite surrounding gas or Molecular Chaos), puts in evidence the different time-
correlations of the two baths [26]: eventually, the observer, by means of some "filter",
should be able to sort out their different contributions Q_b and Q_g. A model where this
separation is explicit has been discussed in Sect. 4.3-a discussion on its time-reversal
properties is given below, in Sect. 5.2.4.

5.2.2 Linear Response

If one applies to Eq. (4.32) an external time-dependent external force $F(t)$, it appears
that

$$\langle \delta V(t) \rangle = \langle V(t) \rangle_{F(t)} - \langle V(t) \rangle_{F \equiv 0} = \int_{-\infty}^{t} ds\, R(t - s) F(s) \tag{5.19}$$

with $R(s)$ the so-called "response". Equation (5.19) is a direct consequence of the
linearity of the Langevin equation. In general, for non-linear equations, one may
still use (5.19) to define the response function, neglecting terms of higher order in

$F(t)$, which makes sense if $F(t)$ is small enough. Obviously, in the impulsive case $F(t) = F_0\delta(t)$ one immediately has

$$R(t) = \frac{\langle \delta V(t) \rangle}{F_0}. \tag{5.20}$$

It is straigthforward to realize that, in our case:

$$R(t) = \frac{C(t)}{T_{tr}}, \tag{5.21}$$

which, integrated in time, gives

$$\int_0^\infty R(t) = \frac{1}{T_{tr}} D \tag{5.22}$$

which is known as Einstein relation. The integral on the left hand side is the so-called mobility: it corresponds to the ratio V_∞/F_0, when V_∞ is the asymptotic velocity reached by the intruder when a constant force F_0 is applied from time 0 (i.e. $F(t) = F_0\theta(t)$).

The Einstein relation is a particular case of a more general theorem, the so-called Equilibrium Fluctuation-Dissipation relation (EFDR), which is valid for small perturbations of a system at equilibrium, i.e. a system with stationary probability in phase space given by $\sim\exp(-\beta H(\mathbf{r}, \mathbf{v}))$. In such a system, when the perturbation appears as an additive contribution $-h(t)A(\mathbf{r}, \mathbf{v})$ to the Hamiltonian, it is found for the linear response [12, 18]

$$R_{Oh} = \frac{\langle O(t) \rangle_{h(t)} - \langle O(t) \rangle_{h\equiv 0}}{\delta h} = -\langle O(t)\dot{A}(0) \rangle_{h\equiv 0}. \tag{5.23}$$

This is a fundamental result expressing a deep relation between linear response to a perturbation and correlations measured in the absence of the perturbation. In the last decades a large amount of scientific literature has been devoted to the study of generalization of this relation to non-equilibrium situations [18]. The fact that a tracer in a driven diluted granular fluid satisfies the EFDR has been seen numerically [19, 20] and, very recently, also in experiments [9].

5.2.3 The Granular Motor

As discussed in Sect. 4.4, when the intruder is not isotropic and its anisotropy breaks symmetry with respect to a fixed direction (for instance in the triangle example of

Fig. 4.1), a spontaneous constant force F appears, see for instance Eq. (4.48). This leads to an asymptotic average velocity $\overline{V} = F/M\gamma$.

The time extensive contribution to the entropy production for this system reads

$$\mathscr{W}_t = \log \frac{P(\Omega_0^t)}{P(\mathscr{I}\Omega_0^t)} \approx \frac{T_r - T}{4T_r T}\sqrt{2\pi T}\gamma[X(t) - X(0)]. \qquad (5.24)$$

Note that, on average, $\mathscr{W}_t > 0$ since $T_r < T$ and $X(t) < X(0)$. It is always $\langle|X(t) - X(0)|\rangle \sim t$ since the ratchet has an average constant velocity $\overline{V}$.

In conclusion the asymmetry unveils the non-equilibrium property of the granular intruder even in the large mass limit. It is interesting at this point to verify that the breakdown of time reversal also breaks the EFDR [32].

It is immediate to see that, for this system, the linear response reads

$$R = \frac{\langle V(t)V(0)\rangle - \overline{V}^2}{T_r}, \qquad (5.25)$$

which is a *violation* of the Einstein relation $R = C(t)/C(0)$ (e.g. Eq. 5.21). As expected, in the absence of detailed balance, the EFDR breaks down. This example is quite simple: indeed the equation for the massive ratchet can be recast in an equation for the variable $z(t) = V(t) - \overline{V}$ which is an equilibrium Langevin equation, for this variable response and correlation are proportional as in the Einstein relation. Anyway, such a re-casting hides out the lack of time-reversal symmetry expressed by relation (5.24): the reason is that the new variable $z(t)$ has not a well defined symmetry with respect to time-reversal ($V(t) - \overline{V}$ goes into $-V(t) - \overline{V}$ when time is inverted). Our analysis in terms of $V(t)$ (and not $z(t)$) is, therefore, the only one consistent and the breakdown of the EFDR is real, even if very simple.

More general out-of-equilibrium Fluctuation-Dissipation relations can be found in the very recent literature, see for instance [1, 3, 17, 18, 28].

5.2.4 Coupling with the Fluid: Non-equilibrium Re-established

In Sect. 4.3, we have discussed a stochastic model for the intruder's dynamics, which is a fair approximation for moderate densities of the fluid. In Eqs. (4.42), the non-Markovian effects due to the interaction between the intruder and the fluid are accounted for by an auxiliary variable $U(t)$. The auxiliary variable represents the *fluctuating* local velocity field near the intruder.

An analysis of the steady state properties of Eqs. (4.42), along the same lines of Sect. 5.1.3, shows that entropy production takes the form [22, 26],

$$\mathscr{W}_t \approx \Gamma_E \left(\frac{1}{T} - \frac{1}{T_b}\right) \int_0^t ds\, V(s)U(s). \qquad (5.26)$$

This functional vanishes exactly in the elastic case, $r = 1$, where equipartition holds, $T = T_b$, and is zero on average in the dilute limit, where $\langle VU \rangle = 0$. Formula (5.26) reveals that the leading source of entropy production is the energy transferred by the "force" $\Gamma_E U$ on the tracer, weighed by the difference between the inverse temperatures of the two "thermostats".

Therefore, to measure entropy production, one needs to measure the fluctuations of U, that is a local average of particles' velocities in the proximities of the intruder. In [26] such a procedure has been carried out and the Fluctuation Relation Eq. (5.8) has been verified in numerical simulations at moderate densities.

A further success of the simple two-variables model in Eq. (4.42) comes with the prediction of the linear response which, at moderate densities, *does not* satisfy the Einstein relation Eq. (5.21). Indeed model (4.42) predicts $C(t)/C(0) = f_C(t)$ and $R(t) = f_R(t)$ with

$$f_{C(R)} = e^{-gt}[\cos(\omega t) + a_{C(R)} \sin(\omega t)]. \tag{5.27}$$

The variables g, ω, a_C and a_R are known algebraic functions of Γ_E, T, Γ', M' and T_b. In particular, the ratio $a_C/a_R = [T - \Omega(T_b - T)]/[T + \Omega(T_b - T)]$, with $\Omega = \Gamma_E/((\Gamma' + \Gamma_E)(\Gamma_E M'/M - \Gamma'))$. Hence, in the elastic ($T \to T_b$) as well as in the dilute limit ($\Gamma' \to \infty$), one gets $a_C = a_R$ and recovers the Einstein relation $C(t)/C(0) = R(t)$. Such predictions have been verified in numerical simulations [26]. Some of the results of [26] are reproduced in Fig. 5.1 which depicts correlation and response functions in a dense case (elastic and inelastic): symbols correspond to numerical data and continuous lines the analytical curves. In the inelastic case, deviations from $R(t) = C(t)/C(0)$ are observed. In the inset of Fig. 5.1 the ratio $R(t)C(0)/C(t)$ is also reported. Very recently an experimental verification of this whole scenario has been obtained by studying the linear response of a rotating probe in a shaken granular medium [9].

Fig. 5.1 Normalized correlation function $C(t)/C(0)$ (*black circles*) and response function $R(t)$ (*red squares*) for $r = 1$ and $r = 0.6$, at $\phi = 0.33$. *Continuous lines* show curves obtained with Eq. (5.27). *Inset* the ratio $R(t)/C(t)$ is reported in the same cases

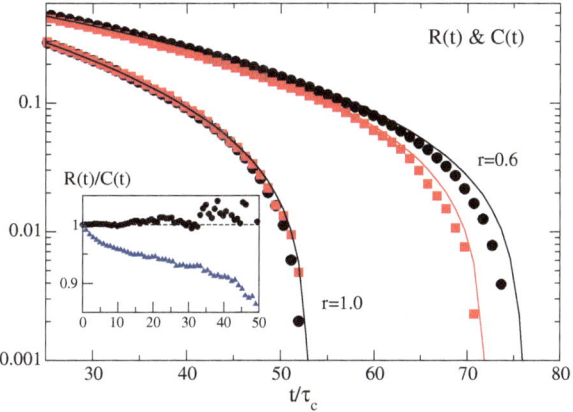

It is important to notice that the main responsibility for the breakdown of the Einstein relation is the coupling between V and U [2], indeed Eq. (5.27) can be expressed in a different way: $R(t) = aC(t) + b\langle V(t)U(0)\rangle$ with $a = [1 - (T - T_b)\Omega_a/\Gamma']$ and $b = (T - T_b)\Omega_b$, where Ω_a and Ω_b are known functions of the parameters. At equilibrium or in the dilute limit the Einstein relation is recovered [19].

To conclude this section, we stress that velocity correlations $\langle V(t)U(t')\rangle$ between the intruder and the surrounding velocity field are responsible for both the violations of the Einstein relation and the appearance of a non-zero entropy production. It must be stressed that *coupling is different from correlation*. Indeed a coupling between V and a local field U certainly exists also for dense fluids *at equilibrium*. In this case, anyway, the two temperatures in Eq. (4.42) are equal and a simple computation shows that $\langle V(t)U(t)\rangle = 0$. The variables are coupled but not correlated.

We also mention that larger violations of Einstein relation can be observed using an intruder with a mass equal or similar to that of other particles [20], with the important difference that in such a case a simple "Langevin-like" model for the intruder's dynamics is not available. In this case the system obeys the full Master Equation (4.1).

5.3 Time-Reversal in Fluctuating Hydrodynamics

The principle learnt above roughly states that a coupling, between variables at different temperatures, is sufficient to break equilibrium. Remarkable consequences are: correlations among variables which are uncorrelated at equilibrium, the appearance of currents or entropy production, the breakdown of the EFDR.

Another interesting situation, in this book, where coupled stochastic equations for different fields appear, is fluctuating hydrodynamics, discussed in Sect. 3.4. In particular, it has been mentioned that the linear approximation is already capable of useful qualitative predictions for correlations, i.e. structure factors.

Let us consider the spatially homogeneous models discussed in this book, which are the homogeneous cooling under suitable rescaling, see Sect. 2.3.3, and the "bulk"-driving, see Sect. 2.3.5. In both cases, linearization around the homogeneous state, a suitable choice of noises (see Sect. 3.4) and a space-Fourier transform, lead to the following Langevin equation:

$$\dot{\tilde{a}}_i(k) = A_{ij}(k)\tilde{a}_j(k) + \xi_i(k), \qquad (5.28)$$

where the specific form of the dynamic matrix A and of the noise amplitudes depend on the kind of thermostat, see [11]. In Eq. (5.28), in two spatial dimensions $(d = 2)$, for each wavevector k, $\tilde{a}$ is a four-dimensional complex field vector $\tilde{a} = (\delta n(k), \delta T(k), u(k), v(k))$. By u and v we mean the longitudinal and transverse velocity field, respectively.

Static and dynamic structure factors in the steady state $\langle \tilde{a}_i(k, t)\tilde{a}_j^*(k, 0)\rangle$ can be computed from Eq. (5.28). They give information on correlation among hydrodynamic fields. Some of these correlations vanish at equilibrium, which corresponds to the elastic limit. A noteworthy example is given by correlations in the velocity field, which are absent at equilibrium, but become observable in the granular (inelastic) case, as discussed in Sect. 3.4, in particular for the transverse modes we have seen the simple expression Eq. (3.76). The complete study of correlations for the homogeneous cooling state has been done in [30], for the bulk-driving without viscosity in [31], and for the case of finite viscosity in [11].

A study of entropy production rate for such three models [10] reveals a common structure to all such models:

$$J_t(k) = h(k)\Re[\langle \delta n(k, t)\dot{T}^*(k, t)\rangle], \qquad (5.29)$$

with $h(k)$ a complicate function of k and of the parameters of the system which vanishes in the elastic case. In particular in the bulk-driving model with finite viscosity, where a bath-temperature T_b is defined, it is seen that $h(k) \propto T_b - T$. Equation (5.29) identifies the correlation among density and temperature time-derivative as the source of violation of detailed balance. The study of Eq. (5.29) as a function of mode number k for the three models [10], shows that:

- for bulk-driving with finite viscosity, J_t has a maximum at a finite k, while goes to zero in both limits $k \to \infty$ and $k \to 0$: this is consistent with the picture that both large scales and small scales are at equilibrium with temperature T_b and T respectively;
- for bulk-driving without viscosity, J_t takes a finite value for $k \to 0$ and decreases toward 0 as $k \to \infty$: in the absence of a macroscopic damping, the system is out of equilibrium even at large scales;
- for the homogeneous cooling, J_t diverges for $k \to 0$, signaling the instability at a finite small k.

A study of *non-linear* fluctuating hydrodynamics is still lacking and certainly deserves attention in the future.

References

1. Baiesi, M., Maes, C., Wynants, B.: Fluctuations and response of nonequilibrium states. Phys. Rev. Lett. **103**, 010602 (2009)
2. Crisanti, A., Puglisi, A., Villamaina, D.: Non-equilibrium and information: the role of cross-correlations. Phys. Rev. E **85**:061127 (2012)
3. Crisanti, A., Ritort, F.: Violation of the fluctuation-dissipation theorem in glassy systems: basic notions and the numerical evidence. J. Phys. A **36**, R181 (2003)
4. de Groot, S.R., Mazur, P.: Non-equilibrium Thermodynamics. Dover Publications, New York (1984)
5. Derrida, B., Brunet, É.: Le mouvement brownien et le théorème de fluctuation-dissipation. In: Einstein aujourd'hui, p. 203. EDP (2005)
6. Evans, D.J., Cohen, E.G.D., Morriss, G.P.: Probability of second law violations in shearing steady flows. Phys. Rev. Lett. **71**, 2401 (1993)

7. Evans, D.J., Searles, D.J.: Equilibrium microstates which generate second law violating steady states. Phys. Rev. E **50**, 1645 (1994)
8. Gallavotti, G., Cohen, E.G.D.: Dynamical ensembles in stationary states. J. Stat. Phys. **80**, 931 (1995)
9. Gnoli, A., Puglisi, A., Sarracino, A., Vulpiani, A.: Nonequilibrium brownian motion beyond the effective temperature. Plos One **9**, e93720 (2014)
10. Gradenigo, G., Puglisi, A., Sarracino, A.: Entropy production in non-equilibrium fluctuating hydrodynamics. J. Chem. Phys. **137**, 014509 (2012)
11. Gradenigo, G., Sarracino, A., Villamaina, D., Puglisi, A.: Fluctuating hydrodynamics and correlation lengths in a driven granular fluid. J. Stat. Mech. P08017 (2011)
12. Kubo, R., Toda, M., Hashitsume, N.: Statistical Physics II: Nonequilibrium Stastical Mechanics. Springer, Berlin (1991)
13. Kurchan, J.: Fluctuation theorem for stochastic dynamics. J. Phys. A **31**, 3719 (1998)
14. Langevin, P.: Sur la theorie du mouvement brownien. C. R. Acad. Sci. (Paris), **146**, 530 (1908). Translated in Am. J. Phys. **65**, 1079 (1997)
15. Lebowitz, J.L., Spohn, H.: A Gallavotti-Cohen-type symmetry in the large deviation functional for stochastic dynamics. J. Stat. Phys. **95**, 333 (1999)
16. Liphardt, J., Dumont, S., Smith, S.B., Tinoco, I., Bustamante, C.: Equilibrium information from nonequilibrium measurements in an experimental test of Jarzynski's equality. Science **296**, 1832 (2002)
17. Lippiello, E., Corberi, F., Zannetti, M.: Off-equilibrium generalization of the fluctuation dissipation theorem for ising spins and measurement of the linear response function. Phys. Rev. E **71**, 036104 (2005)
18. Marconi, U.M.B., Puglisi, A., Rondoni, L., Vulpiani, A.: Fluctuation-dissipation: response theory in statistical physics. Phys. Rep. **461**, 111 (2008)
19. Puglisi, A., Baldassarri, A., Loreto, V.: Fluctuation-dissipation relations in driven granular gases. Phys. Rev. E **66**, 061305 (2002)
20. Puglisi, A., Baldassarri, A., Vulpiani, A.: Violations of the Einstein relation in granular fluids: the role of correlations. J. Stat. Mech. P08016 (2007)
21. Puglisi, A., Rondoni, L., Vulpiani, A.: Relevance of initial and final conditions for the fluctuation relation in markov processes. J. Stat. Mech. P08010 (2006)
22. Puglisi, A., Villamaina, D.: Irreversible effects of memory. Europhys. Lett. **88**, 30004 (2009)
23. Puglisi, A., Visco, P., Trizac, E., van Wijland, F.: Dynamics of a tracer granular particle as a nonequilibrium markov process. Phys. Rev. E **73**, 021301 (2006)
24. Ritort, F.: Nonequilibrium fluctuations in small systems: from physics to biology. Adv. Chem. Phys. **137**, 31 (2008)
25. Sarracino, A., Villamaina, D., Costantini, G., Puglisi, A.: Granular brownian motion. J. Stat. Mech. P04013 (2010)
26. Sarracino, A., Villamaina, D., Gradenigo, G., Puglisi, A.: Irreversible dynamics of a massive intruder in dense granular fluids. Europhys. Lett. **92**, 34001 (2010)
27. Seifert, U.: Stochastic thermodynamics, fluctuation theorems and molecular machines Rep. Prog. Phys. **75**, 126001 (2012)
28. Speck, T., Seifert, U.: Restoring a fluctuation-dissipation theorem in a nonequilibrium steady state. Europhys. Lett. **74**, 391 (2006)
29. Tolman, R.C.: The Principles of Statistical Mechanics. Dover Publications, New York (2003)
30. van Noije, T.C.P., Ernst, M.H., Brito, R., Orza, J.A.G.: Mesoscopic theory of granular fluids. Phys. Rev. Lett. **79**, 411 (1997)
31. van Noije, T.P.C., Ernst, M.H., Trizac, E., Pagonabarraga, I.: Randomly driven granular fluids: large-scale structure. Phys. Rev. E **59**, 4326 (1999)
32. Villamaina, D.: Transport Properties in Non-equilibrium and Anomalous Systems. Springer, Berlin (2013)
33. Visco, P.: Work fluctuations for a Brownian particle between two thermostats. J. Stat. Mech. P06006 (2006)
34. Zamponi, F.: Is it possible to experimentally verify the fluctuation relation? A review of theoretical motivations and numerical evidence. J. Stat. Mech. P02008 (2007)

Conclusion and Perspectives

In the five chapters of this book I traced an itinerary across the wide subject of granular fluids. The leitmotif of this personal walkthrough has been the discussion of *analogies and differences* with respect to molecular systems.

A first analogy is the possibility of using kinetic theory, starting with an adaptation of the Boltzmann equation, to describe the statistical properties of dilute granular models and experiments. Analogies also include the possibility of a hydrodynamic description, when slow space-time scales can be separated from the fast ones. Such a condition is more common than originally thought: somehow, granular hydrodynamics has a wider range of application than what expected from simple estimates. Another clear point of contact with ordinary fluids is the appearance of transitions to ordered structures (patterns) which are well described by hydrodynamic instabilities. Last but not least, the dynamics of tracers dispersed in granular fluids is, for many purposes, difficult to be distinguished from tracer's normal diffusion, e.g. that of colloids in molecular solvents. Violations of the equilibrium fluctuation-dissipation relations, in many situations, can be small and not easily detected.

The most evident difference is the inherent *non-equilibrium* nature of granular fluids: even a spatially homogeneous granular gas is constantly dissipating kinetic energy and, to be stationary, needs energy injection from external sources. As a result, common phenomena which, in molecular fluids, may arise only under an external forcing (which introduces spatial currents), in granular fluids occur spontaneously: shear and cluster instabilities are a fundamental example. A second striking difference is the *small* number of elementary constituents, which is several orders of magnitude lower than the typical size of molecular systems. This is reflected upon the magnitude of fluctuations, which are much more relevant than in ordinary fluids. As a matter of facts, fluctuations in a granular system can rarely be ignored. This makes granular fluids an ideal benchmark for the many theories on non-equilibrium fluctuations which enriched statistical physics in the last two decades.

From the perspective outlined here, one may highlight some problems deserving further attention in the next future. Only a few attempts have been done toward microscopic derivations of *granular noise*, e.g. for hydrodynamic fluctuations or tracer's diffusion. Such problem is an occasion to develop new statistical tools, or

© The Author(s) 2015
A. Puglisi, *Transport and Fluctuations in Granular Fluids*,
SpringerBriefs in Physics, DOI 10.1007/978-3-319-10286-3

refine the existing ones, such as fluctuating Boltzmann equations and techniques based upon projection operators. Large deviation theory offers a promising route, but it must be adapted to the peculiarities of granular hydrodynamics considering, e.g., the importance of inertia and the necessity to cope with vectorial fields. Another important open problem is that of the fluid-solid (ordered or glassy) transition and coexistence, and the role played in it by fluctuations. Such an issue is strictly related to that of *dense granular fluids*, where the Molecular Chaos assumption fails because of frequent recollisions. A promising starting point in this direction seems to be the study of tracer's dynamics, which displays memory effects as the host fluid's density grows. A fascinating connection could be conceived with microrheology in driven (e.g. sheared) jammed granular systems. Last but not least, an almost unexplored territory is that of the foundations of granular hydrodynamics, which relies upon the fast *relaxation to local equilibrium* of many degrees of freedom. The H-theorem, which is only valid for conservative interactions, should be replaced here by some new guiding principle (see Sect. 2.3.6 for a recent attempt). Hydrodynamics foundations also need the correct choice of *relevant fields*: for instance the choice of temperature can be rigorously justified only in the limit of elastic collisions. When fast and slow scales are not separated by many orders of magnitudes, the choice of fields is non-trivial and could hold surprises.

Appendix A
Expansion of the First Two Moments of the Transition Rates for Large Mass of the Tracer

For larger generality, in this Appendix I discuss the case where the gas surrounding the intruder may have a non-zero average $\mathbf{u}$:

$$p(\mathbf{v}) = \frac{1}{\sqrt{(2\pi T/m)^d}} \exp\left[-\frac{m(\mathbf{v} - \mathbf{u})^2}{2T}\right] \qquad (A.1)$$

which is a task involving only the definition of new shifted variables

$$\mathbf{c} = \mathbf{V} - \mathbf{u} \qquad (A.2)$$
$$\mathbf{c}' = \mathbf{V}' - \mathbf{u}. \qquad (A.3)$$

We are interested in computing

$$
\begin{aligned}
D_i^{(1)}(\mathbf{V}) &= \int d\mathbf{V}'(V_i' - V_i) W_{tr}(\mathbf{V}'|\mathbf{V}) \\
&= \int d\mathbf{c}'(c_i' - c_i)\chi \frac{1}{\sqrt{2\pi T/mk(\varepsilon)^2}} \\
&\quad \times \exp\left\{-m\left[c_\sigma' + (k(\varepsilon) - 1)c_\sigma\right]^2 / (2Tk(\varepsilon)^2)\right\}.
\end{aligned}
\qquad (A.4)
$$

In order to perform the integral, we make the following change of variables (see Fig. A.1 for an example)

$$
\begin{aligned}
c_\sigma &= c_x \frac{c_x' - c_x}{\sqrt{(c_x' - c_x)^2 + (c_y' - c_y)^2}} + c_y \frac{c_y' - c_y}{\sqrt{(c_x' - c_x)^2 + (c_y' - c_y)^2}} \\
c_\sigma' &= c_x' \frac{c_x' - c_x}{\sqrt{(c_x' - c_x)^2 + (c_y' - c_y)^2}} + c_y' \frac{c_y' - c_y}{\sqrt{(c_x' - c_x)^2 + (c_y' - c_y)^2}}
\end{aligned}
\qquad (A.5)
$$

© The Author(s) 2015
A. Puglisi, *Transport and Fluctuations in Granular Fluids*,
SpringerBriefs in Physics, DOI 10.1007/978-3-319-10286-3

Fig. A.1 An example for the change of variables $(c'_x, c'_y) \to (c_\sigma, c'_\sigma)$, introduced in Eq. (A.5). Such change of variable, when inverted, has two possible determinations: in this example both represented vectors $\mathbf{c}'$ yield the same (c_σ, c'_σ)

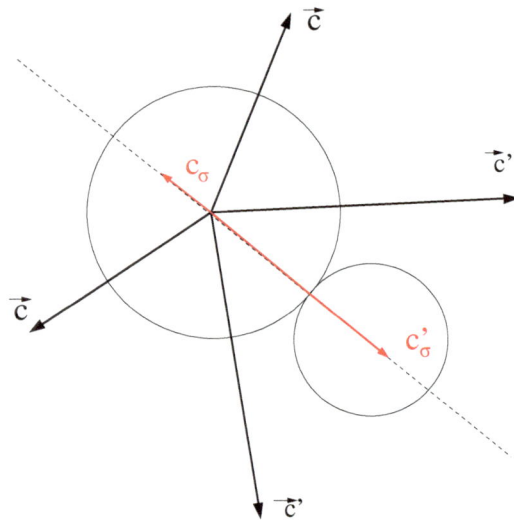

which implies

$$d\mathbf{c}' = dc'_x dc'_y \to dc_\sigma dc'_\sigma |J|, \tag{A.6}$$

where

$$|J| = \frac{|c'_\sigma - c_\sigma|}{\sqrt{c_x^2 + c_y^2 - c_\sigma^2}} \Theta(c_x^2 + c_y^2 - c_\sigma^2) \tag{A.7}$$

is the Jacobian of the transformation. The collision rate is then

$$r(\mathbf{V}) = \chi \sqrt{\frac{\pi}{2T/m}} e^{-\frac{mc^2}{4T}} \left[(c^2 + 2T/m) I_0 \left(\frac{mc^2}{4T} \right) + c^2 I_1 \left(\frac{mc^2}{4T} \right) \right], \tag{A.8}$$

where $I_n(x)$ are the modified Bessel functions. For $D_i^{(1)}$ we can write

$$
\begin{aligned}
D_i^{(1)}(\mathbf{V}) &= \chi \int_{-\infty}^{+\infty} dc_\sigma \int_{c_\sigma}^{\infty} dc'_\sigma (c'_i - c_i) |J| \frac{1}{\sqrt{2\pi T/m} k(\varepsilon)^2} \\
&\quad \times \exp\left\{ -m \left[c'_\sigma + (k(\varepsilon) - 1) c_\sigma \right]^2 / (2Tk(\varepsilon)^2) \right\} \\
&= \chi \int_{-c}^{+c} dc_\sigma \int_{c_\sigma}^{\infty} dc'_\sigma (c'_i - c_i) \frac{c'_\sigma - c_\sigma}{\sqrt{c^2 - c_\sigma^2}} \\
&\quad \times \frac{1}{\sqrt{2\pi T/m} k(\varepsilon)^2} \exp\left\{ -m \left[c'_\sigma + (k(\varepsilon) - 1) c_\sigma \right]^2 / (2Tk(\varepsilon)^2) \right\} \quad \text{(A.9)}
\end{aligned}
$$

where we have enforced the constraint of the theta function, namely $c_\sigma \in (-c, +c)$, with $c = \sqrt{c_x^2 + c_y^2}$. Notice that the integral in dc'_σ is lower bounded by the condition $c'_\sigma \geq c_\sigma$ which follows from the definition of c_σ. In order to compute the integral, we have to invert the transformation (A.5). That yields two determinations for the variables c'_x and c'_y (see Fig. A.1)

$$(A) \begin{cases} c'_x - c_x = \frac{c'_\sigma - c_\sigma}{c^2}\left(c_\sigma c_x + c_y \mathrm{Sign}(c_x)\sqrt{c^2 - c_\sigma^2}\right) \\ c'_y - c_y = \frac{c'_\sigma - c_\sigma}{c^2}\left(c_\sigma c_y - c_x \mathrm{Sign}(c_x)\sqrt{c^2 - c_\sigma^2}\right) \end{cases}$$

$$(B) \begin{cases} c'_x - c_x = \frac{c'_\sigma - c_\sigma}{c^2}\left(c_\sigma c_x - c_y \mathrm{Sign}(c_x)\sqrt{c^2 - c_\sigma^2}\right) \\ c'_y - c_y = \frac{c'_\sigma - c_\sigma}{c^2}\left(c_\sigma c_y + c_x \mathrm{Sign}(c_x)\sqrt{c^2 - c_\sigma^2}\right) \end{cases}$$

Then the integral (A.9) can be written as

$$D_x^{(1)}(\mathbf{V}) = \chi \int_{-c}^{c} dc_\sigma \int_{c_\sigma}^{\infty} dc'_\sigma \left[(c'_x - c_x)^{(A)} + (c'_x - c_x)^{(B)}\right] |J|$$
$$\times \frac{1}{\sqrt{2\pi T/m}k(\varepsilon)^2} \exp\left\{-m\left[c'_\sigma + (k(\varepsilon) - 1)c_\sigma\right]^2/(2Tk(\varepsilon)^2)\right\},$$

$$(A.10)$$

yielding

$$D_x^{(1)} = -\frac{2}{3}\chi k(\varepsilon)\sqrt{\frac{m\pi}{2T}}c_x e^{-\frac{mc^2}{4T}}\left[(c^2 + 3T/m)I_0(\frac{mc^2}{4T}) + (c^2 + T/m)I_1(\frac{mc^2}{4T})\right],$$

$$D_y^{(1)} = -\frac{2}{3}\chi k(\varepsilon)\sqrt{\frac{m\pi}{2T}}c_y e^{-\frac{mc^2}{4T}}\left[(c^2 + 3T/m)I_0(\frac{mc^2}{4T}) + (c^2 + T/m)I_1(\frac{mc^2}{4T})\right].$$

$$(A.11)$$

Analogously, for the coefficients $D_{ij}^{(2)}$ one obtains

$$D_{xx}^{(2)}(\mathbf{V}) = \frac{1}{2}\chi \int_{-c}^{c} dc_\sigma \int_{c_\sigma}^{\infty} dc'_\sigma \left[\left((c'_x - c_x)^{(A)}\right)^2 + \left((c'_x - c_x)^{(B)}\right)^2\right] |J|$$
$$\times \frac{1}{\sqrt{2\pi T/m}k(\varepsilon)^2} \exp\left\{-m\left[c'_\sigma + (k(\varepsilon) - 1)c_\sigma\right]^2/(2Tk(\varepsilon)^2)\right\}$$
$$= \frac{1}{2}\chi \frac{k(\varepsilon)^2}{15}\sqrt{\frac{2m\pi}{T}}e^{-\frac{mc^2}{4T}}$$

$$\times \left\{ \left[c^2(4c_x^2 + c_y^2) + 3T(7c_x^2 + 3c_y^2)/m + 15T^2/m^2 \right] I_0 \left(\frac{mc^2}{4T} \right) \right.$$

$$+ \left[c^2(4c_x^2 + c_y^2) + T(13c_x^2 + 7c_y^2)/m + 3T^2/m^2 \frac{-c_x^2 + c_y^2}{c^2} \right]$$

$$\left. I_1 \left(\frac{mc^2}{4T} \right) \right\}, \tag{A.12}$$

$$D_{xy}^{(2)}(\mathbf{V}) = \frac{1}{2}\chi \int_{-c}^{c} dc_\sigma \int_{c_\sigma}^{\infty} dc'_\sigma \left[(c'_x - c_x)^{(A)}(c'_y - c_y)^{(A)} \right.$$

$$+ (c'_x - c_x)^{(B)}(c'_y - c_y)^{(B)} \bigg] |J|$$

$$\times \frac{1}{\sqrt{2\pi T/m} k(\varepsilon)^2} \exp\left\{ -m \left[c'_\sigma + (k(\varepsilon) - 1)c_\sigma \right]^2 / (2Tk(\varepsilon)^2) \right\}$$

$$= \frac{1}{2}\chi \frac{k(\varepsilon)^2}{5} \sqrt{\frac{2m\pi}{T}} e^{-\frac{mc^2}{4T}} c_x c_y$$

$$\times \left[(c^2 + 4T/m) I_0 \left(\frac{mc^2}{4T} \right) + \frac{c^4 + 2c^2 T/m - 2T^2/m^2}{c^2} I_1 \left(\frac{mc^2}{4T} \right) \right]. \tag{A.13}$$

Then we introduce the rescaled variables

$$q_x = \frac{c_x}{\sqrt{T/m}} \varepsilon^{-1} \qquad q_y = \frac{c_y}{\sqrt{T/m}} \varepsilon^{-1}, \tag{A.14}$$

obtaining

$$D_x^{(1)}(\mathbf{V}) = -\frac{2}{3}\chi \sqrt{\frac{\pi}{2} \frac{T}{m}} q_x k(\varepsilon)\varepsilon e^{-\frac{\varepsilon^2 q^2}{4}} \left[(\varepsilon^2 q^2 + 3) I_0(\frac{\varepsilon^2 q^2}{4}) + (\varepsilon^2 q^2 + 1) I_1(\frac{\varepsilon^2 q^2}{4}) \right],$$

$$D_{xx}^{(2)}(\mathbf{V}) = \frac{1}{2}\chi \frac{1}{15}\sqrt{2\pi} \left(\frac{T}{m} \right)^{3/2} k(\varepsilon)^2 e^{-\frac{\varepsilon^2 q^2}{4}}$$

$$\times \left\{ \left[\varepsilon^4 q^2(4q_x^2 + q_y^2) + 3\varepsilon^2(7q_x^2 + 3q_y^2) + 15 \right] I_0 \left(\frac{\varepsilon^2 q^2}{4} \right) \right.$$

$$+ \left[\varepsilon^4 q^2(4q_x^2 + q_y^2) + \varepsilon^2(13q_x^2 + 7q_y^2) + 3\frac{-q_x^2 + q_y^2}{q^2} \right] I_1 \left(\frac{\varepsilon^2 q^2}{4} \right) \right\}$$

$$D_{xy}^{(2)}(\mathbf{V}) = \frac{1}{2}\chi \frac{1}{5}\sqrt{2\pi} \left(\frac{T}{m} \right)^{3/2} q_x q_y k(\varepsilon)^2 \varepsilon^2 e^{-\frac{\varepsilon^2 q^2}{4}}$$

$$\times \left[(\varepsilon^2 q^2 + 4) I_0 \left(\frac{\varepsilon^2 q^2}{4} \right) + \left(\frac{\varepsilon^4 q^4 + 2\varepsilon^2 q^2 - 2}{\varepsilon^2 q^2} \right) I_1 \left(\frac{\varepsilon^2 q^2}{4} \right) \right]. \tag{A.15}$$

Up to this last results we have not introduced any small ε approximation. The next step consists in assuming that $q \sim \mathcal{O}(1)$ with respect to ε, which is equivalent to assume that $c^2 \sim T/M$: this assumption must be compared to its consequences, in particular to Eq. (4.35). When the assumption is consistent, expanding in ε and using that $I_0(x) \sim 1 + x^2/4$ and $I_1(x) \sim x/2$ for small x, one finds Eq. (4.30).

Index

© The Author(s) 2015
A. Puglisi, *Transport and Fluctuations in Granular Fluids*,
SpringerBriefs in Physics, DOI 10.1007/978-3-319-10286-3